FACE OF SHOP III

慕门而来：商业门面设计 III

深圳市海阅通文化传播有限公司　编著

门，面之所在。
facade， where image hides

PREFACE

This is an age which values beauty and appearance so much, businesses could be much more successfull with a pretty "face". As small as clothes, jewelry, cosmetics, red wines, coffees, breads and electric products, or even as large as vehicles, shopping malls, furnishing stores, hospitals, all these are in need of façades—fantastic faces and their designers. We are now in a big-bang days, with a life show us beauty in every place in any time.

The fact tells us that some times a simple pattern with fuzzy words can be the universal language. Could you guess what good these stores offer, by looking at the strange logos and signs on the façade? Lights, colors, accessories, showcase and shop planning, all these elements are made full use to create splendid façade. Façade is not only a space division, but also a direction of human manner, as well as with philosophical meaning and life accumulation.

Face of shop: commercial façade design III keeps the main core of the former two books, to exam the façade design with a wider view. Besides the commercial character, façade also reflects the aesthetics and ideology in a certain age. The statue of façade records the spiritual and material life changes in a country or a state. So we can say, façade is a culture carrier. Especially in nowadays when various cultures meet each others, cultural exchanges become an unavoidable common phenomenon. In such a circumstance, thoughts gather and crush, inspiring a series of exciting design works. No matter the designs designed by foreign designers, or those native designers go abroad; no matter the native styles or the other styles chosen by designers, they can not be the objects been criticized, because all these facts are some kinds of cultural collision forms. What is more, this kind of designs would be more and more popular thanks to the internationalization; thus turns into a new internationalism style.

There are many native projects in this book, as well as overseas works, even top designers with high reputation like Kengo Kuma contributes his masterpieces. Among which, there are many designs with combination of regious and styles, presenting a multi-cultural product to the readers and let them feel the "impossible" and "unbelievable" clearly.

Surrounded by various cultures, China should create a design brand of its own in order to be outstanding, by developing design into creative cultural industry and making cultures to be the soul of design flowing on each project. This book interpreted design from the level of cultural collision, lighten Chinese design industry with the soul of design.

在这个靠脸吃饭的年代，没有一张颜值极高的"脸面"都不好意思做生意。小到服装、珠宝、化妆品、红酒、咖啡、面包、电子产品，大到汽车、购物中心、家居卖场、医院，都已经离不开门面，离不开让空间拥有一张迷人脸庞的门面设计师。我们生活在大爆炸的时代，让我们感受生活处处带给我们的美！

事实告诉我们，有时不需要太多繁复的文字，简单的图案就是世界通用的语言。看看这些门面上千奇百怪的标志，猜猜它们出售的是什么商品、什么灯光、色彩、搭配、橱窗、店铺规划，统统用起来！门面，不仅是空间的划分，也是人们行为的向导，它不仅具有哲学意味，同时也有生活的积淀。

《慕门而来：商业门面设计Ⅲ》继承了前两部的精神，又以更广阔的视野来审视门面的设计。门面，它除了具有商业用途之外，还深刻地反映了一个时代的审美情趣和意识形态。门面的形态，记录着一个国家和地区思想精神和物质生活的变迁。可以说，门面是一个文化载体。尤其是在文化多元碰撞的今天，文化的交流成为人们不能回避的普遍现实。尤其是设计领域的思想碰撞，产生出的灵感形成了一个个令人振奋的作品。无论是本土的设计采用外来的设计师，还是本地的设计师走出国门；无论是设计师采用本地风格，还是采用其他风格，这些都不应成为批判的对象，因为这本身就是文化碰撞的一种形式。而且随着国际化的深入，这种形式的设计会越来越多，逐渐形成一种新的"国际主义"风格。

本书不仅采用了大量本土的设计，也选用了很多国外的案例，其中包括隈研吾这样具有很高声望的国际设计大师的作品。作品具有地域和风格交融性，给读者展现出一个多元文化的产物，让人们感受以往认为的诸多"不可能"和"想不到"。

在多元文化的今天，中国要形成自己的设计品牌，一定是将设计发展成为创意文化产业，让文化成为设计的灵魂，流动在每一个作品之中。本书就是从文化碰撞的层面来解读设计，让设计师品味设计的灵魂。

资深设计评论人　王颖超

CONTENTS

006-061 弄潮专卖：服饰
Fashion Store: Clothing

062-177 乐享美食：中西餐饮
Delicacy Food: Restaurant

178-211 水文化：咖啡、茶、酒
Water Culture: Coffee, Tea, Wine

212-239 快乐养生：会所
Happy Regimen:Club

240-251 花想容：美容美发
Pretty Beauty: Beauty Salon

252-265 珠宝配饰：黄金玉器
Jewelry Accessories:Gold,Jade

266-311 爱家：艺术家居
Love Home: Artistic Furnishing

312-323 聚焦：售楼处
Focus:Sales Office

324-351 其他：书店、诊所
Other Else: Bookstore, Clinic

专卖店

水文化

其他

珠宝配饰

爱家

乐享美食

售楼处

花想容

快乐养生

弄潮专卖：服饰
Fashion Store: Clothing

Estnation Nagoya
Estnation 名古屋专卖店

Design agency: MOMENT
Designer: Hisaaki Hirawata , Tomohiro Watabe
Location: Nagoya,Japan
Area: 387m²
Photography: Nacasa & Partners Inc.

设计单位：Moment 设计事务所
设计师：Hisaaki Hirawata , Tomohiro Watabe
项目地点：日本名古屋
项目面积：387 平方米
摄影：Nacasa & Partners Inc. 摄影工作室

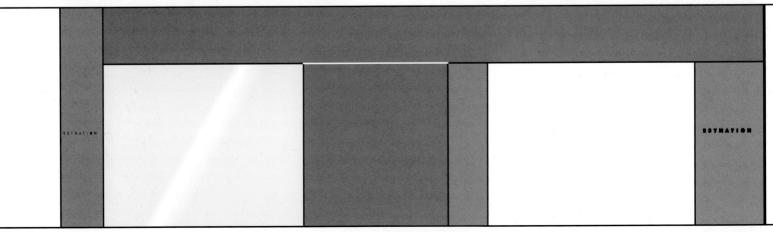

SHANG XIA
Store in Paris
上下品牌巴黎零售店

Design agency: Kengo Kuma and Associates
Designer: Kengo Kuma
Location: Paris, France

设计单位：隈研吾建筑都市设计事务所
设计师：隈研吾
项目地点：法国巴黎

Karl Lagerfeld Store, Beijing INTIME Lotte

卡尔·拉格斐乐天银泰百货店

Design agency: Plajer & Franz Studio
Location: Beijing
Area: 263m²
Photography: Karl Lagerfeld Ltd. China

设计单位：Plajer & Franz 工作室
项目地点：北京
项目面积：263 平方米
摄影：中国卡尔·拉格斐品牌

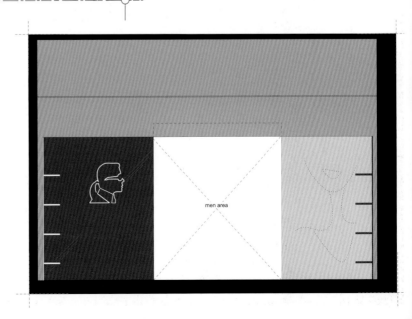

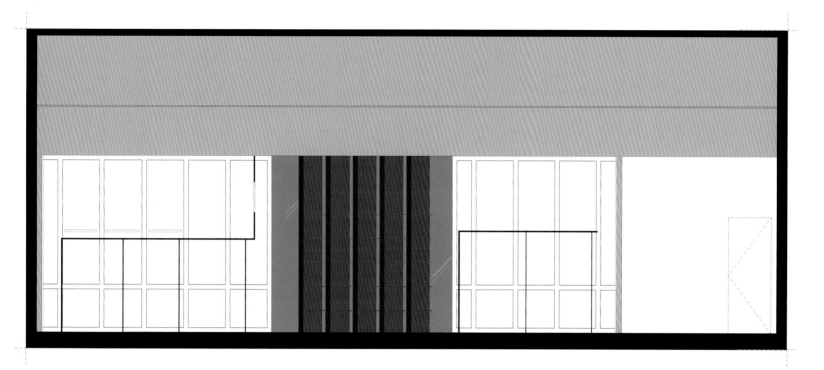

NICKIE in Lishui

丽水 NICKIE 童装店

Design agency: SAKO Architects
Designer: Keiichiro SAKO, Shuhei Aoyama, Sara Aghajani
Location: Lishui, Zhejiang
Area: 78m²
Photography: Ruijing Photo

设计单位：SAKO 建筑设计工社
设计师：Keiichiro SAKO, Shuhei Aoyama, Sara Aghajani
项目地点：浙江丽水
项目面积：78 平方米
摄影：Ruijing Photo 摄影工作室

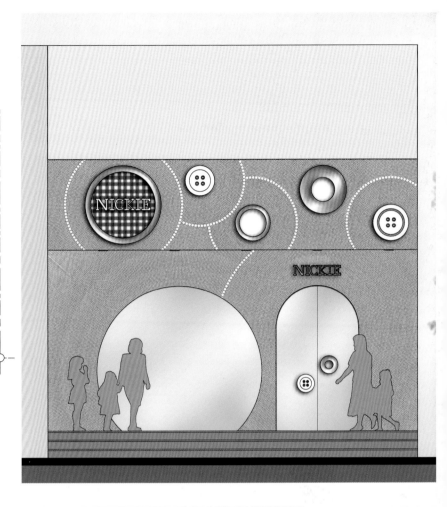

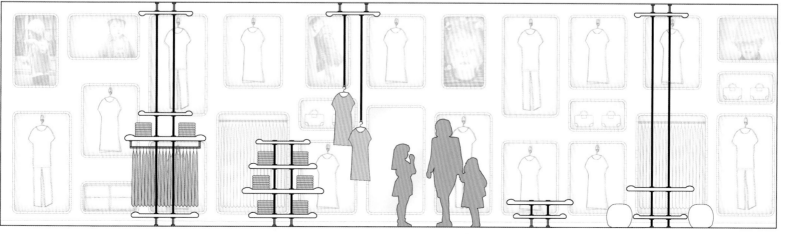

016 FACE OF SHOP

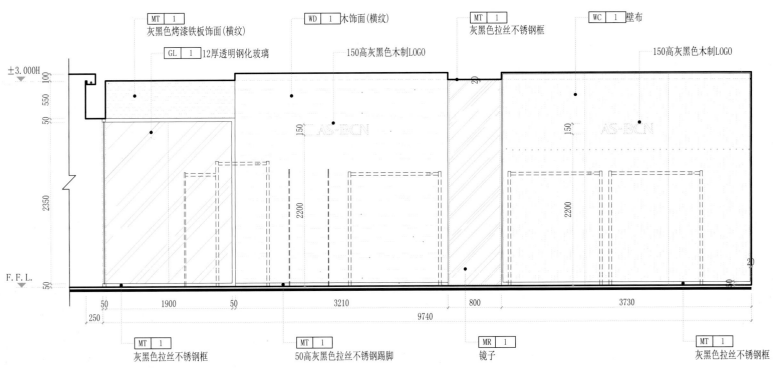

AS-BCN Fashion Store
欧思卡时尚女装店

Design agency: Shanghai Window Design Co., Ltd. 设计单位：上海尚窗室内设计有限公司
Designer: Cheryl Lee 设计师：李秀儿
Location: Shanghai 项目地点：上海
Area: 90m² 项目面积：90 平方米

BAO BAO ISSEY MIYAKE

Bao Bao 三宅一生潮袋专卖店

Design agency: Moment	设计单位：Moment 设计事务所
Designer: Hisaaki Hirawata, Tomohiro Watabe	设计师：Hisaaki Hirawata, Tomohiro Watabe
Location: Marunouchi, Japan	项目地点：日本丸之内
Area: 34m²	项目面积：34 平方米
Photography: Fumio Araki	摄影：Fumio Araki

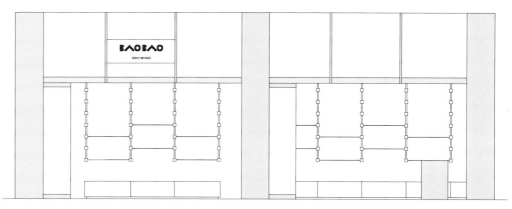

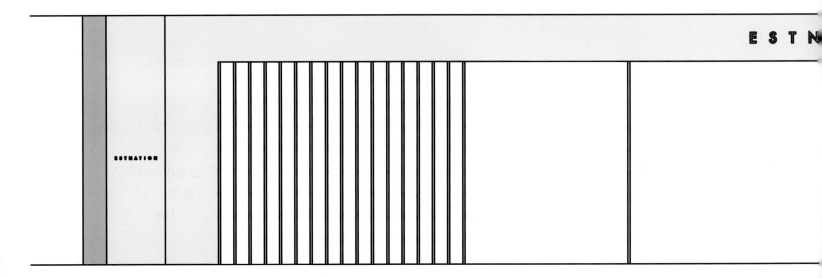

Estnation Tamagawa

Estnation 多摩川专卖店

Design agency: Moment
Designer: Hisaaki Hirawata, Tomohiro Watabe
Location: Tamagawa, Japan
Area: 348m²
Photography: Fumio Araki

设计单位：Moment 设计事务所
设计师：Hisaaki Hirawata, Tomohiro Watabe
项目地点：日本多摩川
项目面积：348 平方米
摄影：Fumio Araki

Brooks Brothers
Fashion Boutique

布克兄弟高级服装店

Design agency: Stefano Tordiglione Design Ltd.
Designer: Stefano Tordiglione
Location: IFC, Hong Kong
Area: 80m²
Photography: Edmon Leong

设计单位：Stefano Tordiglione Design Ltd.
设计师：Stefano Tordiglione
项目地点：香港国际金融中心
项目面积：80 平方米
摄影：Edmon Leong

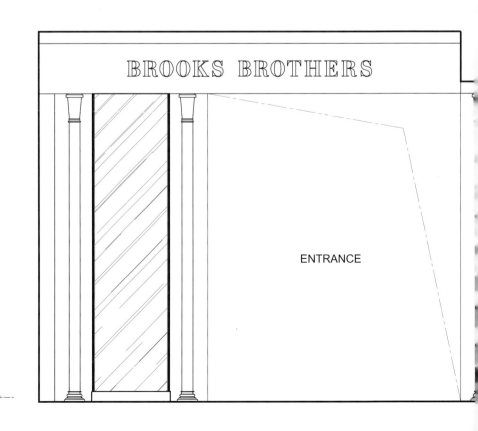

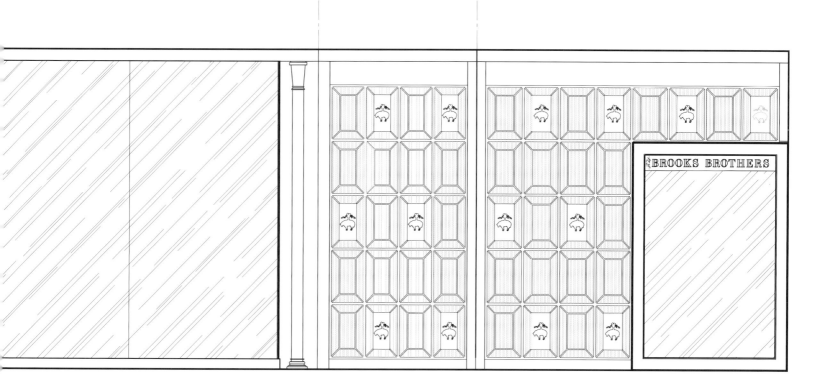

Max Mara Rome
罗马麦丝玛拉专卖店

Design agency: Duccio Grassi Architects
Location : Rome, Italy
Area: 713m²
Photography: Andrea Martiradonna

设计单位：杜乔·格拉西建筑事务所
项目地点：意大利罗马
项目面积：713 平方米
摄影：安德里亚·马蒂阿杜娜

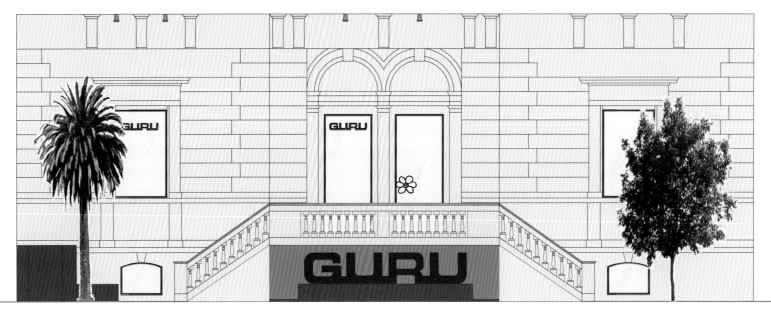

Guru Palermo

巴勒莫 Guru 潮牌店

Design agency: Duccio Grassi Architects
Designer: Duccio Grassi
Location: Palermo, Italy
Area: 93m²
Photography: Andrea Martiradonna

设计单位：杜乔·格拉西建筑事务所
设计师：杜乔·格拉西
项目地点：意大利巴勒莫
项目面积：93 平方米
摄影：安德里亚·马蒂阿杜娜

Babywalz
Brand Exhibition Hall
贝沃兹母婴品牌展示厅

Design agency: Shanghai Window Design Co., Ltd.
Designer: Cheryl Lee
Location: Wenzhou, Zhejiang
Area: 560m²
Main materials: colorized plastic flooring, fraxinus excelsior, white culture brick, wallpaper, etc.

设计单位：上海尚窗室内设计有限公司
设计师：李秀儿
项目地点：浙江温州
项目面积：560平方米
主要材料：彩色塑胶地板、白腊木、白色文化砖、墙纸等

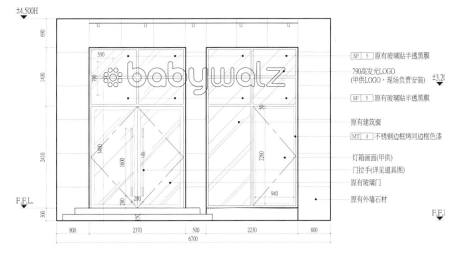

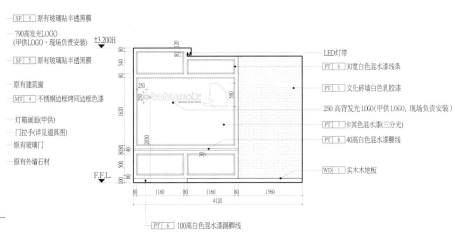

Yumiko Closet
Yumiko 女装店

Design agency: Cynthia's Interior Design Studio
Location: Taiwan
Area: 50m²

设计单位：张馨室内设计事务所
项目地点：台湾
项目面积：50 平方米

Giuseppe men's store

乔治白男装专卖店

Design agency: Shanghai Window Design Co., Ltd.
Designer: Cheryl Lee
Location: Wenzhou, Zhejiang
Area: 268m²

设计单位：上海尚窗室内设计有限公司
设计师：李秀儿
项目地点：浙江温州
项目面积：268 平方米

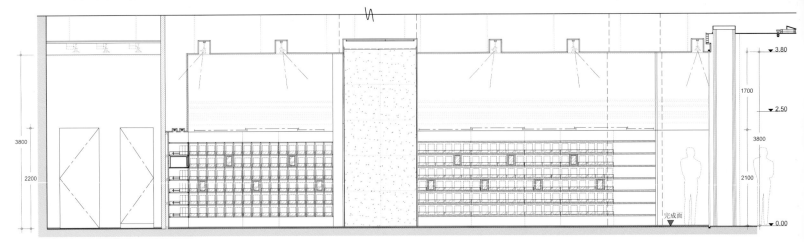

UM Top Fashion Men's Underwear Brand Shop
UM 高端男性内衣综合品牌店

Design agency: AS Design Service Limited
Designer: Four Lau, Sam Sum
Location: Shenzhen, Guangdong
Area: 126m²
Photography: Sing Studio by Sum Sing

设计公司：AS Design Service Limited
设计师：刘盛科、沈浩梁
项目地点：广东深圳
项目面积：126 平方米
摄影：Sing Studio by Sum Sing

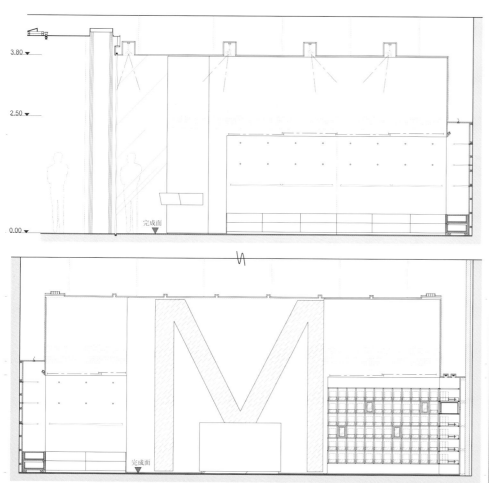

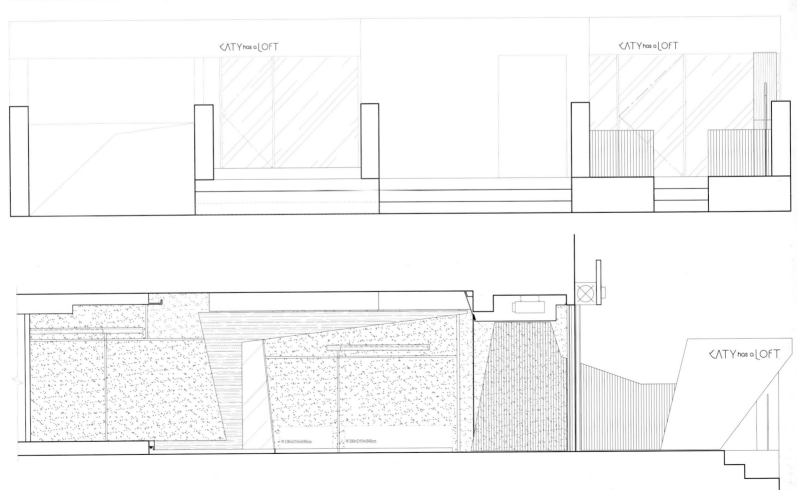

Katy has a Loft
Katy has a Loft 商店

Design agency: Taipei Base Design Center
Designer: Janus Huang, Roy Huang
Location: Taipei, Taiwan
Main materials: marble, dyed veneers, quartz brick, painted glass, metal, plastic flooring tile

设计单位：台北基础设计中心
设计师：黄鹏霖、黄怀德
项目地点：台湾台北
主要材料：大理石、木皮染色、石英砖、烤漆玻璃、金属、塑料地砖

VTwo Shop
VTwo 时尚专卖店

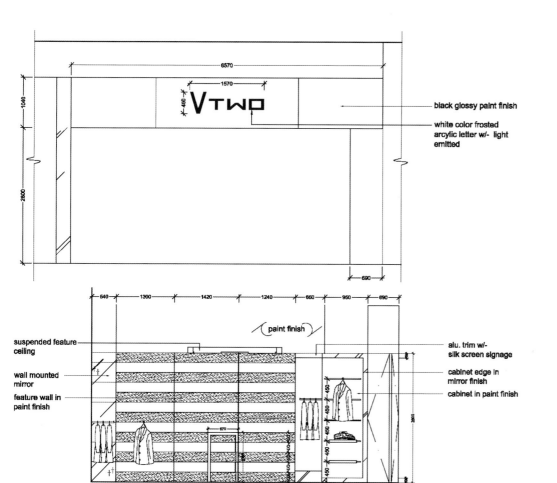

Design agency: Tiko Interiors Ltd.
Designer: Matthew Tong
Location: Hong Kong
Area: 65m²

设计单位：天皓设计工程有限公司
设计师：唐万强
项目地点：香港
项目面积：65 平方米

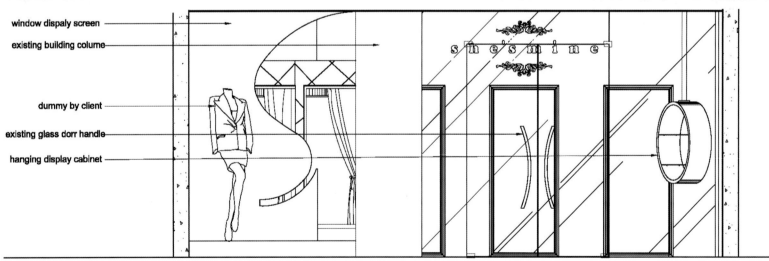

She's Mine Shop

She's Mine 女装店

Design agency: Tiko Interiors Ltd.
Designer: Matthew Tong
Location: Hong Kong
Area: 70m²

设计单位：天皓设计工程有限公司
设计师：唐万强
项目地点：香港
项目面积：70 平方米

Betu Fashion Store
百图时尚专卖店

Design agency: Tiko Interiors Ltd.
Designer: Matthew Tong
Location: Hong Kong
Area: 50m²

设计单位：天皓设计工程有限公司
设计师：唐万强
项目地点：香港
项目面积：50 平方米

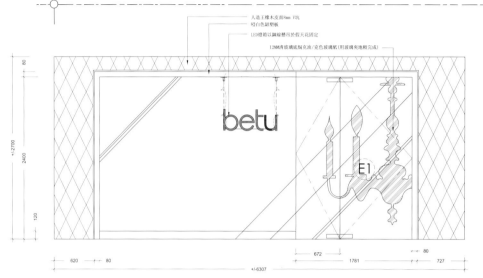

038 FACE OF SHOP

New Look
新风貌服装旗舰店

Design agency: Checkland Kindleysides 设计单位：Checkland Kindleysides 设计顾问公司
Location: Nottingham, UK 项目地点：英国诺丁汉
Area: 2700m² 项目面积：2700 平方米

Uniqlo
优衣库

| Design agency: Checkland Kindleysides | 设计单位：Checkland Kindleysides 设计顾问公司 |
| Location: London, UK | 项目地点：英国伦敦 |

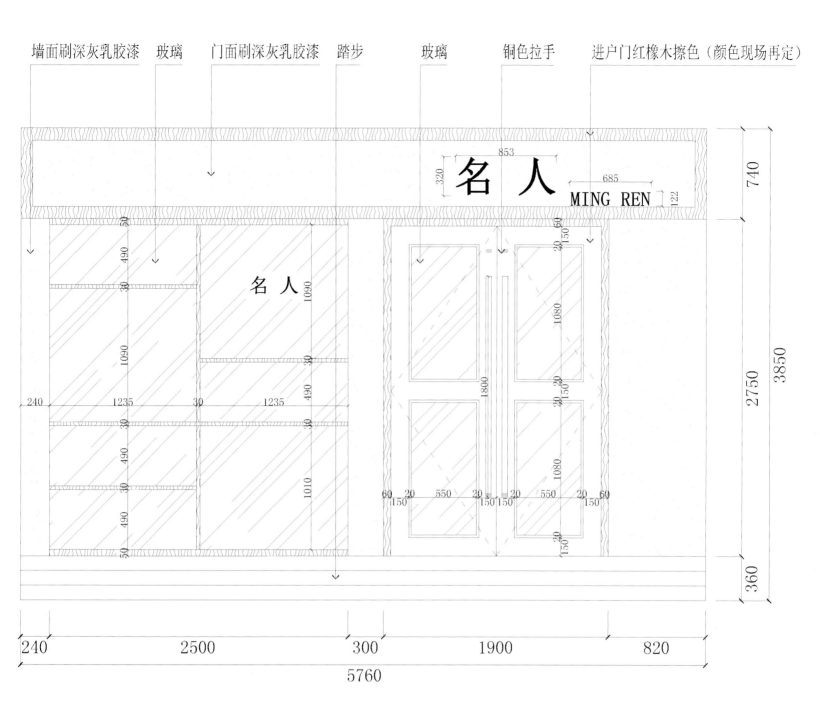

Ming Ren Fashion Store
名人服装店

Design agency: Da Shu Luxuries—You Wei-zhuang Design	设计单位：大墅尚品—由伟壮设计
Location: Changshu, Jiangsu	项目地点：江苏常熟
Area: 200m²	项目面积：200 平方米
Main materials: tile, carpet, latex paint, mirror finished stainless steel, wood, veneer	主要材料：地砖、地毯、乳胶漆、镜面不锈钢、实木、饰面板

New Look Shop

新风貌服装专卖店

Design agency: Checkland Kindleysides
Location: London, UK

设计单位：Checkland Kindleysides 设计顾问公司
项目地点：英国伦敦

Rooms Store
Rooms 时尚女装店

Design agency: AX Design Group
Location: Taipei, Taiwan
Area: 53m²

设计单位：大器联合室内装修设计有限公司
项目地点：台湾台北
项目面积：53 平方米

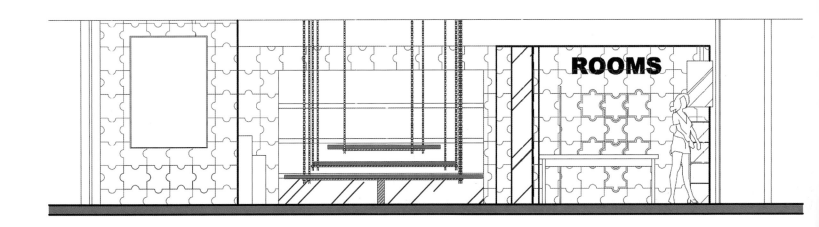

Yuzen
玉臻

Design agency: Yuejie Design Company
Designer: Ming Guo, Bian-ping Yue
Location: Beijing
Area: 140 m²
Photography: Wei Yun
Main materials: black steel, marble, wood flooring, mirror

设计单位：悦界设计
设计师：郭明、岳变萍
项目地点：北京
项目面积：140 平方米
摄影：恽伟
主要材料：黑钢、大理石、木地板、镜片

ALL RIDE Taipei
台北 ALL RIDE 商店

Design agency: AX Design Group
Location: Taipei, Taiwan
Area: 255m²

设计单位：大器联合室内装修设计有限公司
项目地点：台湾台北
项目面积：255 平方米

Ellassay Weekend Fashion Store

歌力思品牌专卖店

Design agency: Shenzhen VMDPE Design Group
Designer: Vinci Chan
Location: Shenzhen, Guangdong
Photography: VMDPE Design Group

设计单位：深圳圆道品牌设计顾问有限公司
设计师：程枫祺
项目地点：广东深圳
摄影：圆道品牌设计顾问有限公司

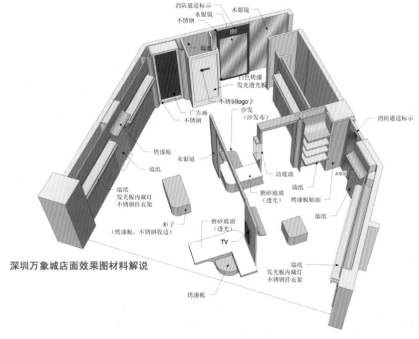

深圳万象城店面效果图材料解说

Hunter's Flagship Store
Hunter 雨靴旗舰店

Design agency: Checkland Kindleysides	设计单位：Checkland Kindleysides 设计顾问公司
Location: London, UK	项目地点：英国伦敦
Area: 235m²	项目面积：235 平方米

Jitaisapphire

Jitaisapphire 实验店

Design agency: Crox International Co., Ltd.
Designer: Tsung-Jen Lin
Location: Taizhou, Jiangsu
Area: 70m²
Photography: Wei-hong Li
Main materials: Cement, stainless steel, LED

设计单位：阔合国际有限公司
设计师：林琮然
项目地点：江苏泰州
项目面积：70平方米
摄影：黎威宏
主要材料：水泥、不锈钢管、LED

PUMA Black Label Store

彪马黑标形象店

Design agency: Plajer & Franz Studio
Location: China
Area: 194m²

设计单位：Plajer & Franz 设计事务所
项目地点：中国
项目面积：194 平方米

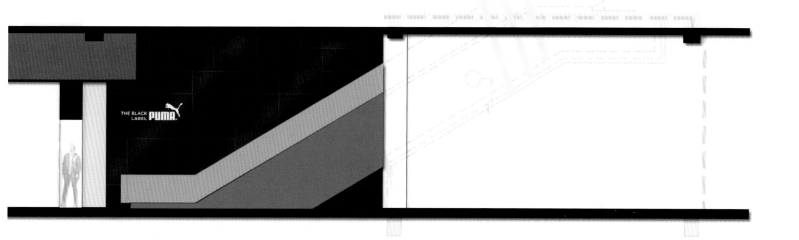

052 FACE OF SHOP

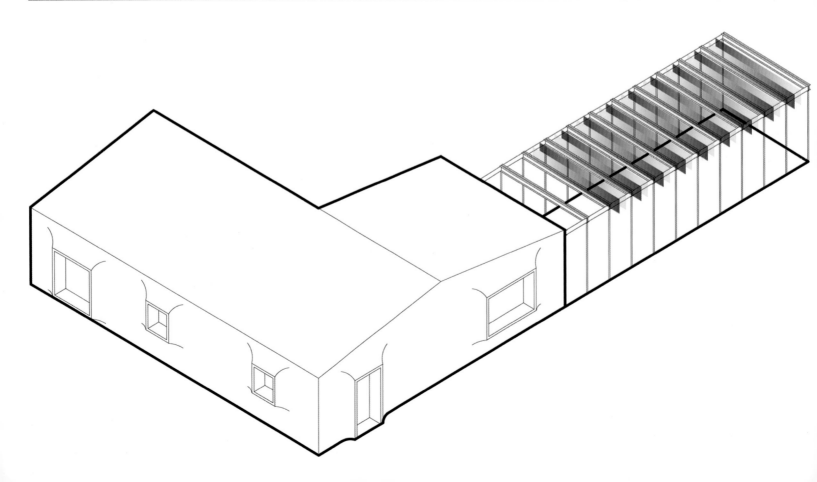

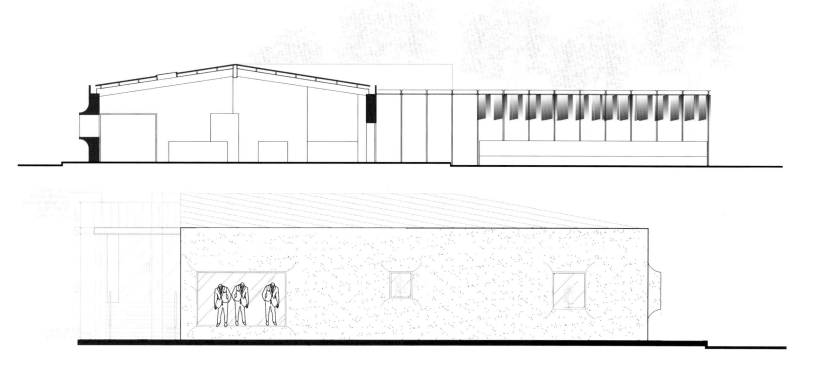

Saul Zona 14
Saul Zona 14

Design agency: Taller KEN	设计单位：Taller KEN 设计事务所
Location: Guatemala City	项目地点：危地马拉
Area: 450m²	项目面积：450 平方米

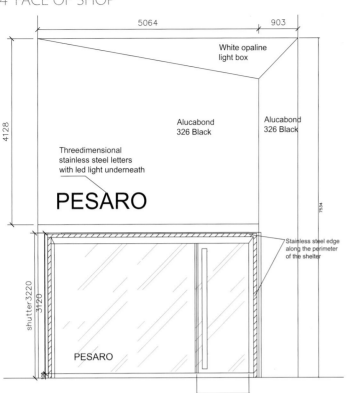

PESARO Retail Branding
黄蕙玲品牌零售店

Design agency: ARBOIT Ltd.
设计单位：艾伯特设计有限公司

Liu Jo Accessories Bari
瑠玖巴里配饰店

Design agency: Fabio Caselli Design
Location: Bari, Italy

设计单位：法比奥·卡塞设计工作室
项目地点：意大利巴里

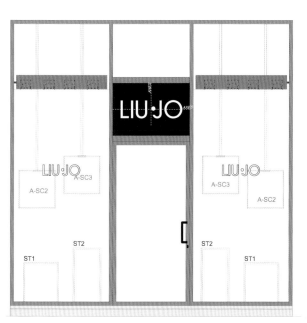

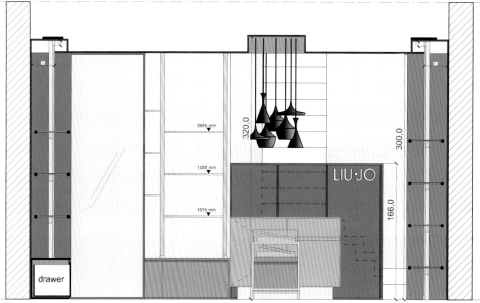

056 FACE OF SHOP

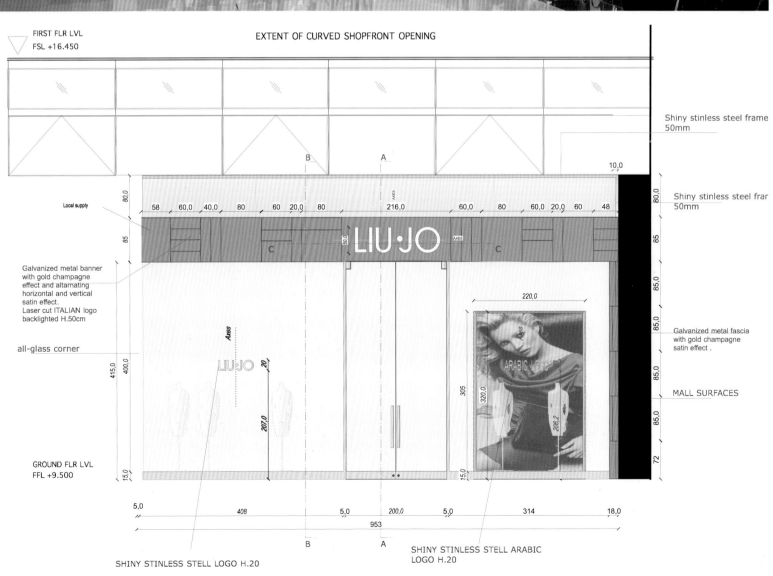

Liu Jo
Dubai Mall
瑠玖迪拜专卖店

Design agency: Fabio Caselli Design
Location: Dubai

设计单位：法比奥·卡塞设计工作室
项目地点：迪拜

Billionaire Couture
亿万富翁服装定制店

Design agency: HEAD Architecture and Design
Location: Macau

设计单位：HEAD 建筑设计有限公司
项目地点：澳门

Tommy Bahama Flagship Store

汤美巴哈马旗舰店

Design agency: HEAD Architecture and Design
Location: Hong Kong

设计单位：HEAD 建筑设计有限公司
项目地点：香港

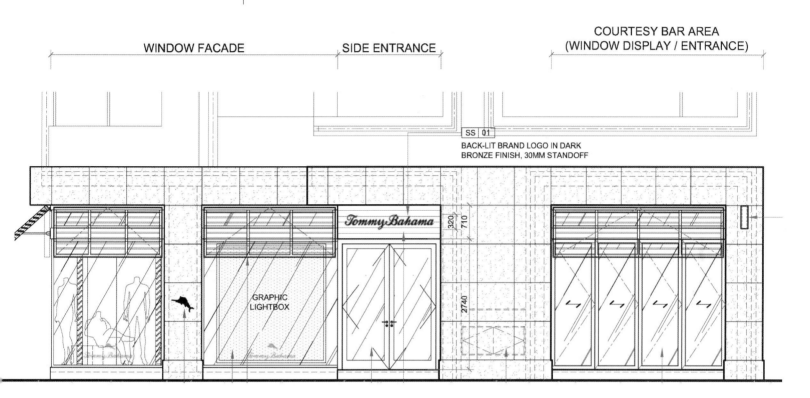

BGilio Store
比丽雅箱包店

Design agency: Tiko Interiors Ltd. 设计单位：天皓设计工程有限公司
Designer: Matthew Tong 设计师：唐万强
Location: Hong Kong 项目地点：香港
Area: 50m² 项目面积：50平方米

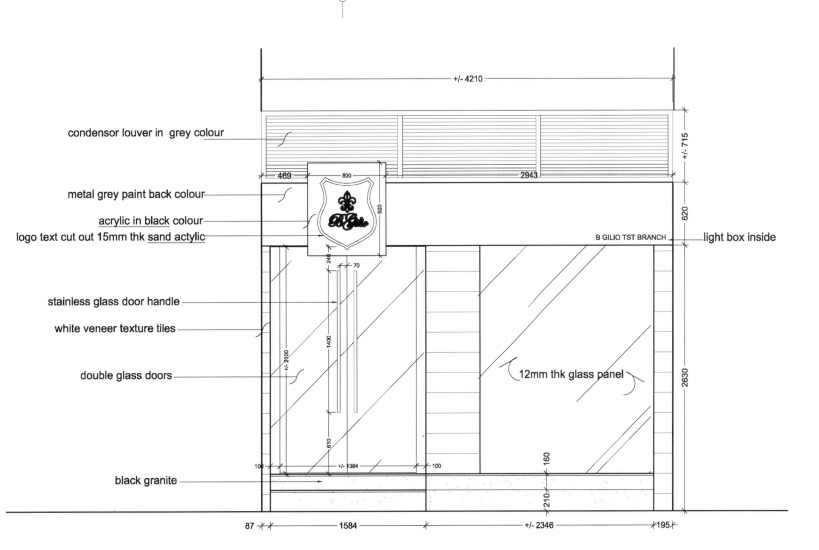

乐享美食：中西餐饮
Delicacy Food: Restaurant

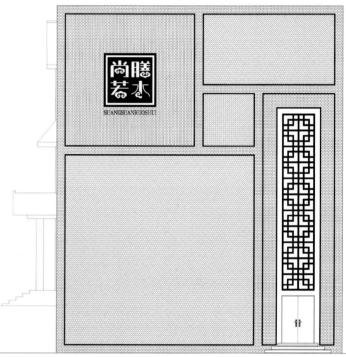

ShangShanRuoShui Restaurant

尚膳若水美食会馆

Design agency: Three Architecture & Interior Design
Designer: Fan Wang, Yuan-chao Wang, Ke Qiu
Location: Shandong
Area: 660m²

设计单位：思锐空间设计有限公司
设计师：王凡、王远超、邱可
项目地点：山东
项目面积：660 平方米

Ding Xiang Creative Cuisine

丁香创意中国菜餐厅

Design agency: Three Architecture & Interior Design
Designer: Fan Wang, Yuan-chao Wang
Location: Shandong
Area: 1680m²

设计单位：思锐空间设计有限公司
设计师：王凡、王远超
项目地点：山东
项目面积：1680 平方米

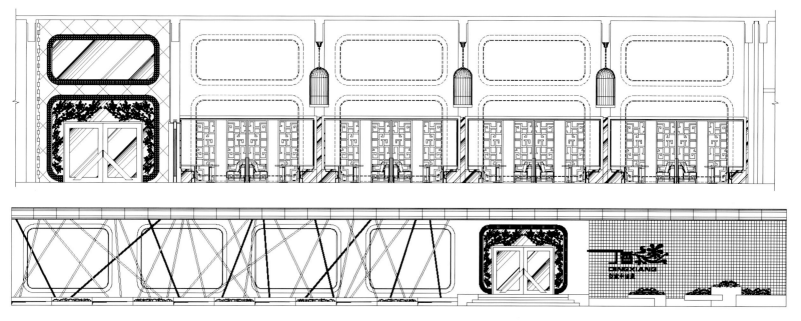

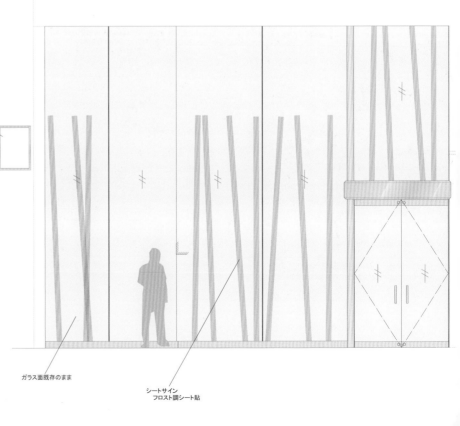

既存電飾看板
シート貼り(切り文字)
(両面共)

ガラス面既存のまま

シートサイン
フロスト調シート貼

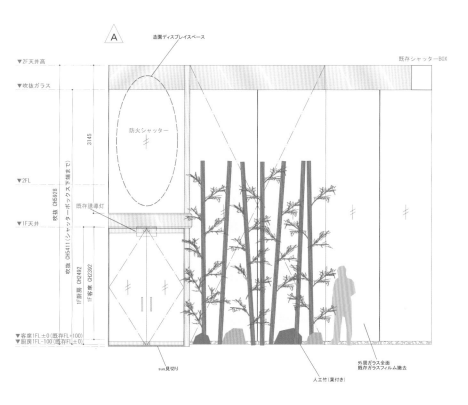
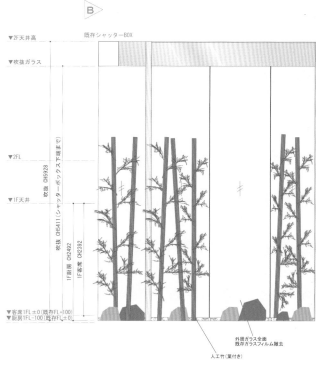

Garaku Steak House
雅罗俱牛排店

Design agency: Matsuya Art Works. Co., Ltd.
Designer: Tetsuya Matsumoto
Location: Kobe, Japan

设计单位：松屋艺术设计有限公司
设计师：Tetsuya Matsumoto
项目地点：日本神户市

Riki-Maru Otsu
大津力丸餐厅

Design agency: Matsuya Art Works. Co., Ltd.
Designer: Tetsuya Matsumoto
Location: Himeji, Japan

设计单位：松屋艺术设计有限公司
设计师：Tetsuya Matsumoto
项目地点：日本姬路市

南側立面図

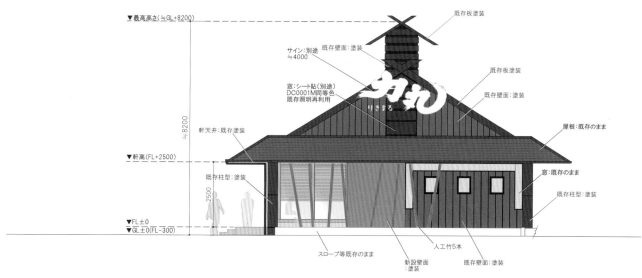

西側立面図

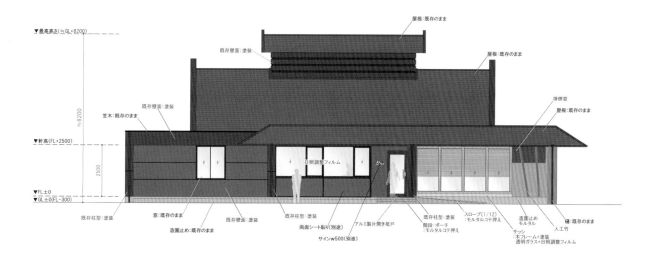

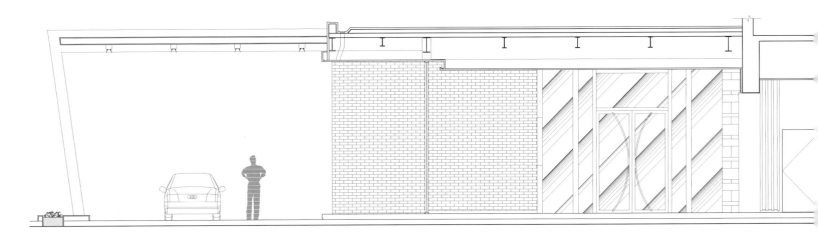

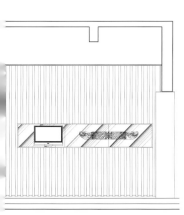

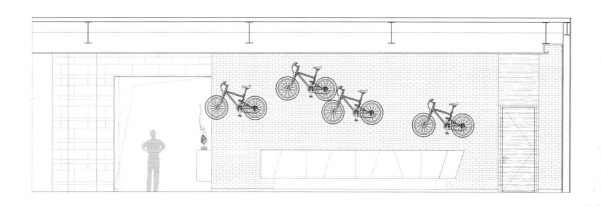

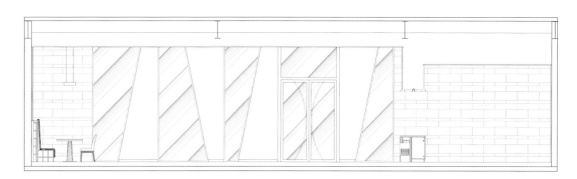

H Hotel

爱驰酒店

Design agency: Three Architecture & Interior Design
Designer: Fan Wang, Yong He, Wei Yue
Location: Shandong
Area: 4000m²

设计单位：思锐空间设计有限公司
设计师：王凡、何勇、岳伟
项目地点：山东
项目面积：4000 平方米

Light House
晓确幸

Design agency: 1/10 Concept Programming Interior
Designer: Tracy Jen
Location: Taipei, Taiwan
Area: 160m²
Photography: Zhen-yu Lu
Main materials: wooden flooring, rustic tile, white glazed tile, stucco washing finished, mirror

设计单位：十分之一设计事业有限公司
设计师：任萃
项目地点：台湾台北
项目面积：160 平方米
摄影：卢震宇
主要材料：耐磨木地板、复古磁砖、白色亮釉面磁砖、洗石子、明镜

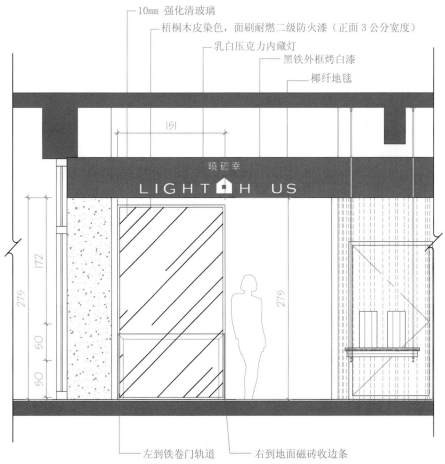

Rocho Creative Restaurant
六潮创意中国菜馆

Design agency: DOLONG Design	设计单位：董龙设计
Photography: Jin Xiao-wen Space Photos	摄影：金啸文空间摄影
Area: 1200m²	项目面积：1200 平方米
Main materials: concrete, imperial kiln brick, flooring, artistic steel net	主要材料：清水混凝土、金砖、地板、艺术钢网

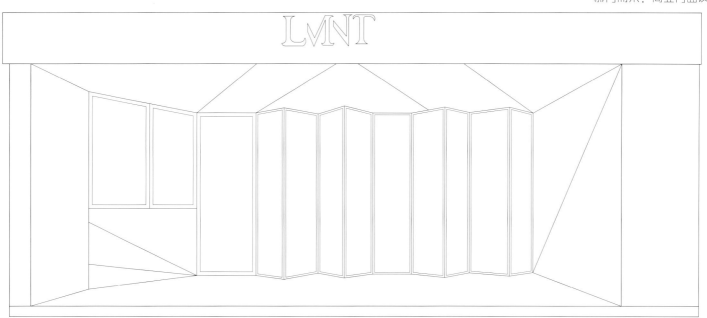

LMNT
LMNT 元素

Design agency: 1/10 Concept Programming Interior
Designer: Tracy Jen
Location: Taipei, Taiwan
Area: 120m²
Photography: Zhen-yu Lu
Main materials: beige travertine, carrara marble, gray mirror, sandblasting sticker, forging iron, wood, white artificial stone

设计单位：十分之一设计事业有限公司
设计师：任萃
项目地点：台湾台北
项目面积：120 平方米
摄影：卢震宇
主要材料：米黄洞石、卡拉拉大理石、灰镜、喷砂卡典贴纸、锻造黑铁、实木、白色人造石

Raki Creative Restaurant
萝喜创意料理餐厅

Design agency: XYI Design Consulting Co., Ltd.
Designer: Mac Huang
Location: Qingdao, Shandong
Area: 400m²
Photography: Ji-shou Wang
Main materials: black iron, mosaic, white oak, gray granite, mirror, iron net, orange acryl

设计单位：隐巷设计顾问有限公司
设计师：黄士华
项目地点：山东青岛
项目面积：400 平方米
摄影：王基守
主要材料：黑铁、马赛克、白橡木、灰色凿面花岗岩、明镜、铁网、橘色压克力

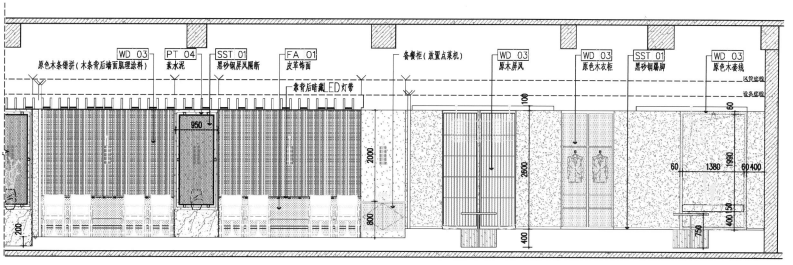

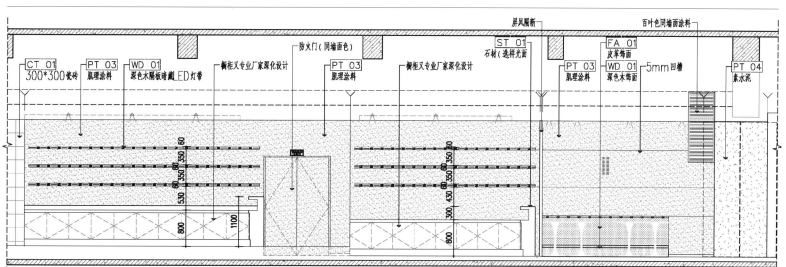

A Kingdom of Wine

煮酒论三国

Design agency: Jiangsu Nantong XingYuTian Design & Decoration Engineering Co., Ltd.; ShiZi ChuPin Design Studio
Designer: Xiao-wei Shi, Wei-duo Kong
Location: Suzhou, Jiangsu
Area: 100m²
Main materials: stone, wood, wallpaper

设计单位：江苏南通行于天装潢设计工程有限公司、石子出品高端设计事务所
设计师：石小伟、孔魏躲
项目地点：江苏苏州
项目面积：100平方米
主要材料：石材、木材、墙纸

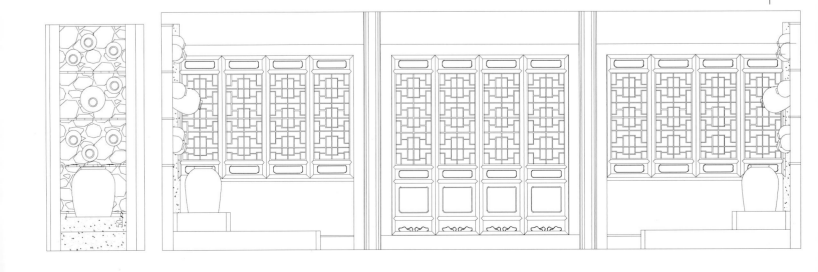

Jianghai Food Museum

江海美食博物馆

Design agency: Jiangsu Nantong XingYuTian Design & Decoration Engineering Co., Ltd.; ShiZi ChuPin Design Studio
Designer: Xiao-wei Shi, Wei-duo Kong
Location: Nantong, Jiangsu
Area: 1000m²
Main materials: wood, bluestone, stone, wallpaper

设计单位：江苏南通行于天装潢设计工程有限公司、石子出品高端设计事务所
设计师：石小伟、孔魏躲
项目地点：江苏南通
项目面积：1000平方米
主要材料：木材、青砖、石材、墙纸

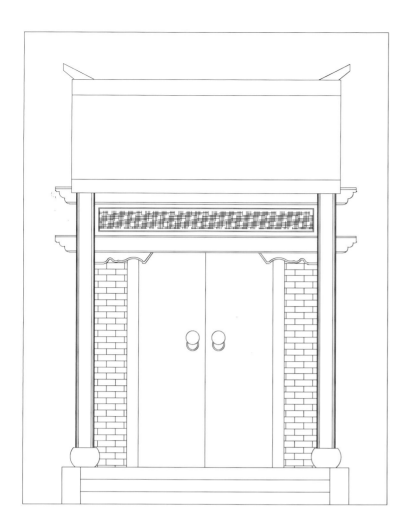
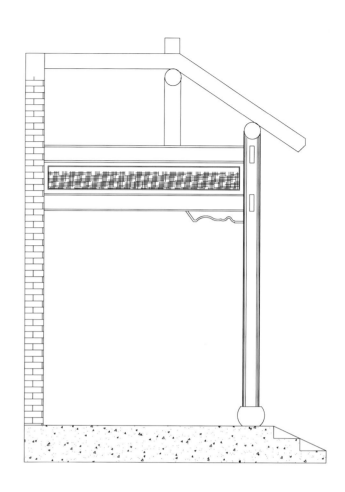

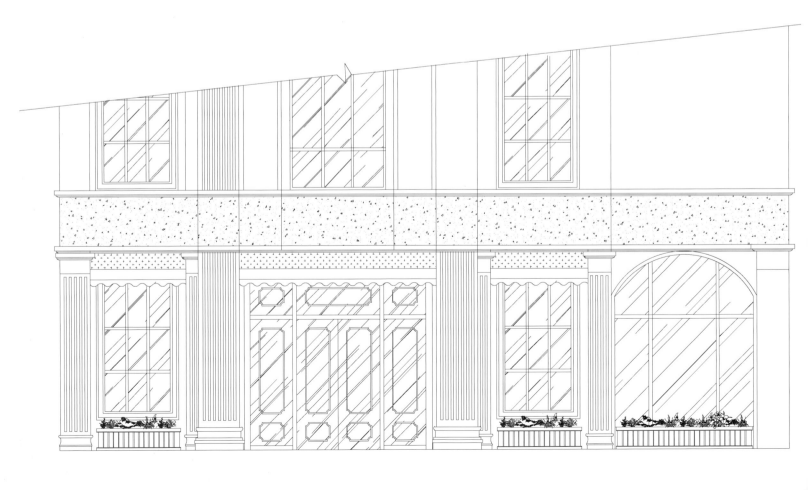

ShangGuan XiaoChu
上官小厨

Design agency: Jiangsu Nantong XingYuTian Design & Decoration Engineering Co., Ltd.; ShiZi ChuPin Design Studio
Designer: Xiao-wei Shi, Wei-duo Kong
Location: Nantong, Jiangsu
Area: 800m²
Main materials: stone, wood, wallpaper

设计单位：江苏南通行于天装潢设计工程有限公司、石子出品高端设计事务所
设计师：石小伟、孔魏躲
项目地点：江苏南通
项目面积：800 平方米
主要材料：石材、木材、墙纸

Shang Chuan Island Restaurant

尚川岛餐厅

Design agency: Shenzhen Yi Ding Design
Designer: Kun Wang
Location: Meizhou, Guangdong

设计单位：深圳市艺鼎装饰设计有限公司
设计师：王锟
项目地址：广东梅州

Casual Paradigm Teahouse

会心典范茶楼

Designer: Yong Wang
Location: Fujian
Area: 185m²
Main materials: rustic tile, floor slab, wallcloth, rosewood, cultural stone

设计师：王勇
项目地址：福建
项目面积：185 平方米
主要材料：仿古砖、水泥板、墙布、花梨木、文化石

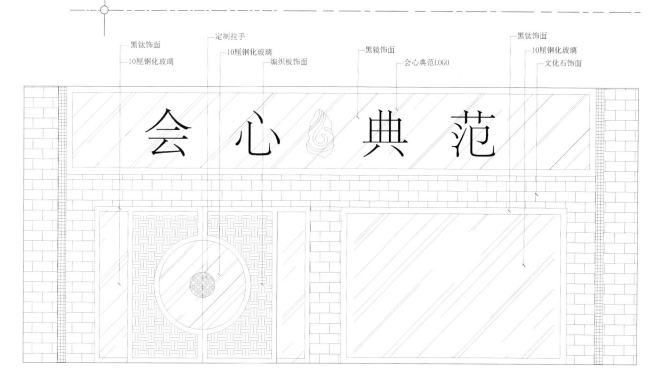

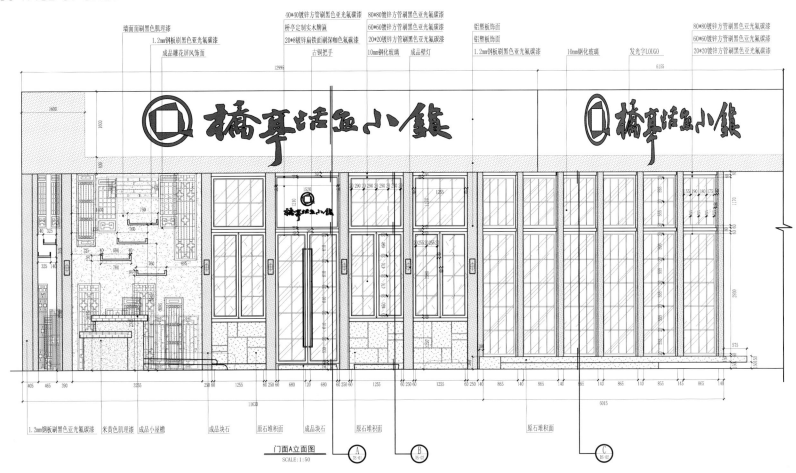

Xihong Road QiaoTing Live Fish Town

西洪路桥亭活鱼小镇

Design agency: Fujian Donny Decorative Engineering & Design Co., Ltd.	设计单位：福建东道建筑装饰设计有限公司
Designer: Chuan-dao Li, Xin-feng Zheng, Li-hui Chen, Hai-ping Zhang	设计师：李川道、郑新峰、陈立惠、张海萍
Location: Fuzhou, Fujian	项目地点：福建福州
Area: 400m²	项目面积：400 平方米
Main materials: wood panel, iron board, tile, square steel, cement paint	主要材料：老木板、铁板、花砖、方钢、水泥漆

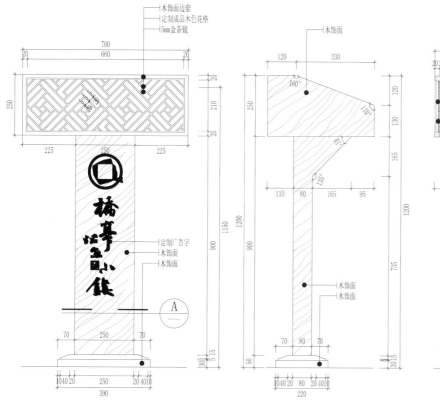
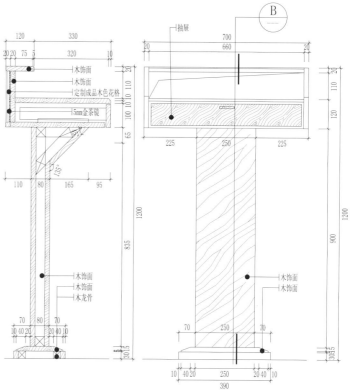

Korean Restaurant

可瑞安韩国料理餐厅

Design agency: DAJ Interior Design Co. Ltd.
Location: Kaohsiung, Taiwan
Area: 130m²
Main materials: fabrics, leather, stone, rustic tile, crystal plates, black mirror, HPL, iron, LED lamp

设计单位：大间空间设计有限公司
项目地址：台湾高雄
项目面积：130平方米
主要材料：裱布、皮革、石材、复古砖、水晶板、墨镜、美耐板、铁件、LED灯

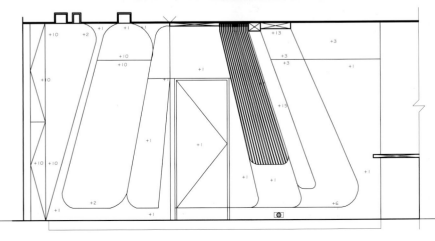

Gushan QiaoTing Live Fish Town

鼓山桥亭活鱼小镇

Design agency: Fujian Donny Decorative Engineering & Design Co., Ltd.
Designer: Chuan-dao Li
Location: Fuzhou, Fujian
Main materials: green brick, bluestone, wood, linen, spray paint, wall paint

设计单位：福建东道建筑装饰设计有限公司
设计师：李川道
项目地点：福建福州
主要材料：青砖、青石板、原木、麻布、喷绘、墙漆

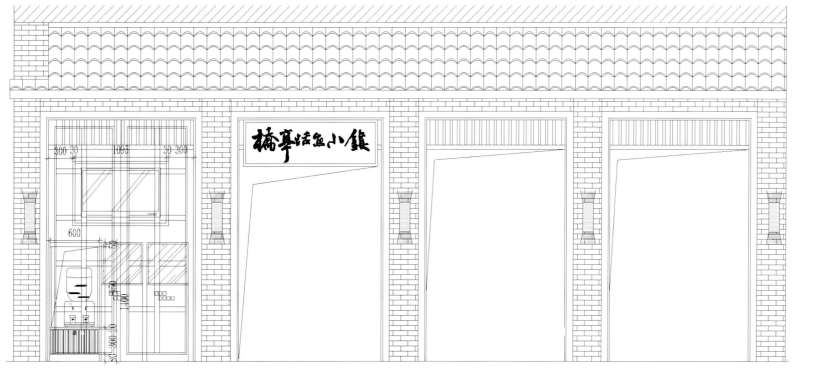

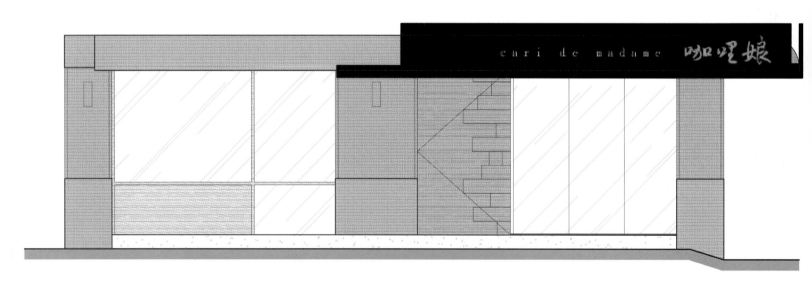

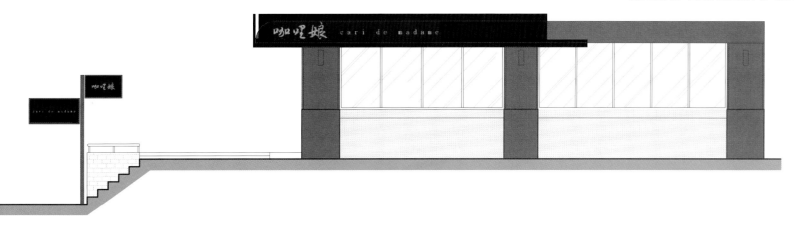

Cari de Madame
咖喱娘

Design agency: Taipei Base Design Center
Designer: Janus Huang, Roy Huang
Location: Taipei, Taiwan

设计单位：台北基础设计中心
设计师：黄鹏霖、黄怀德
项目地点：台湾台北

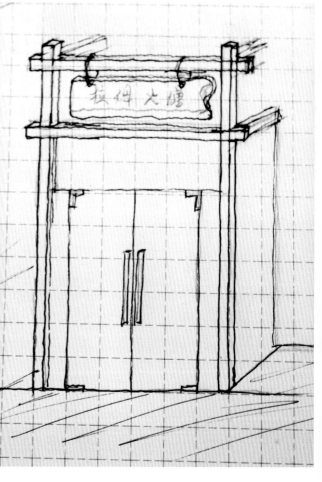

Rahm Pond Restaurant

拉姆火塘餐吧

Design agency: DAS Design Consulting Co., Ltd.
Designer: Jian-ming Li
Location: Kunming, Yunnan
Area: 600m²
Main materials: red brick, original wood, oil painting

设计单位：大森设计顾问有限公司
设计师：李健明
项目地点：云南昆明
项目面积：600 平方米
主要材料：红砖、原木、油画

Pyongyang Arirang Restaurant

平壤阿里郎餐厅

Design agency: Shanghai Scale Art Design Corporation
Designer: Jian-an Lai
Location: Shanghai
Area: 1349m²

设计单位：上海十方圆国际设计工程
设计师：赖建安
项目地点：上海
项目面积：1349 平方米

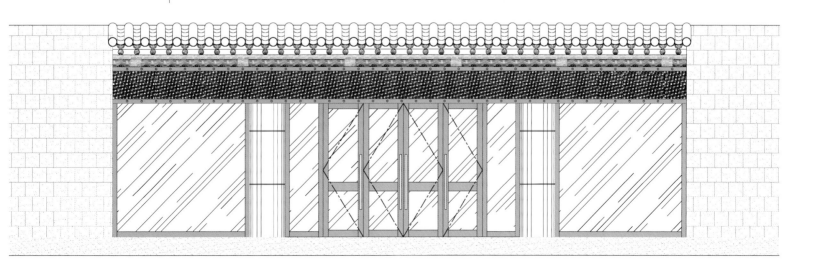

SOGO Korean Dining Restaurant

可瑞安韩国料理餐厅崇光店

Design agency: Dakai Space Design Co., Ltd.
Location: Jungli, Taiwan
Area: 80m²
Main materials: rock slices, mirror plates, powder coating glasses, rustic tile, metal ornamental plates

设计单位：大开空间设计有限公司
项目地点：台湾中坜
项目面积：80 平方米
主要材料：岩片板、镜面板、烤漆玻璃、复古砖、金属美耐板

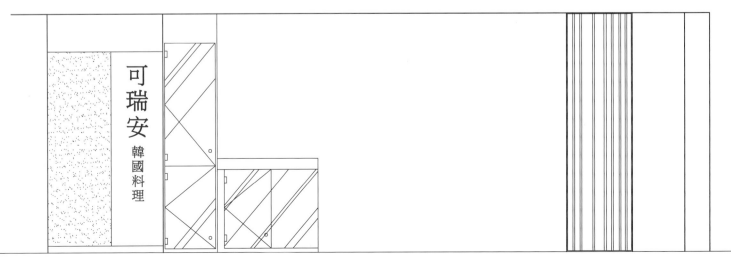

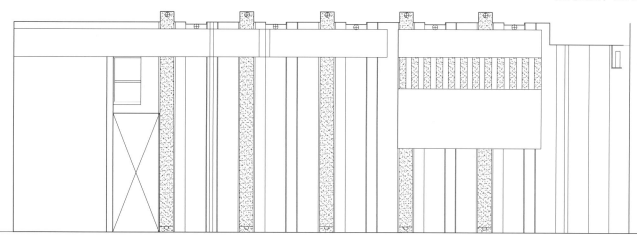

Telepizza

特乐比萨

Design agency: Stone Designs
Location: Madrid, Spain

设计单位：Stone Designs 设计事务所
项目地点：西班牙马德里

LA PASTA
意大利面屋南西店

Design agency: W.C.H. Design Consulting	设计单位：王俊宏建筑设计咨询（上海）有限公司
Designer: Luke Wang	设计师：王俊宏
Location: Taiwan	项目地点：台湾
Area: 290m²	项目面积：290平方米
Photography: KPS Kyle Yu	摄影：KPS 游宏祥

Aburi Yakitori

黑炙餐馆

Design agency: Atelier E Limited
Designer: Nuo Xu
Location: Hong Kong
Area: 120m²

设计单位：Atelier E Limited 设计事务所
设计师：许诺
项目地点：香港
项目面积：120平方米

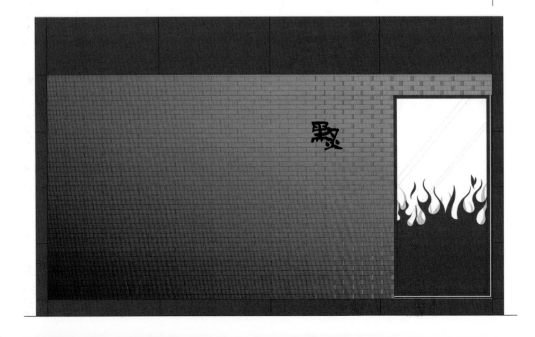

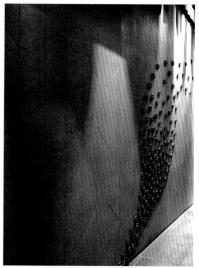
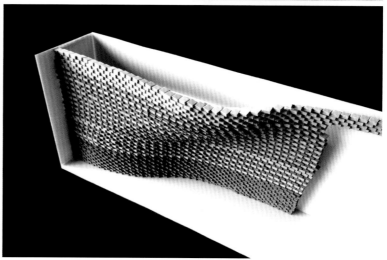

QiFuShen Japanese French Fusion

七福神日本料理店

Design agency: Jiangsu Nantong XingYuTian Design & Decoration Engineering Co., Ltd.; ShiZi ChuPin Design Studio
Designer: Wei-duo Kong, Xiao-wei Shi
Location: Nantong, Jiangsu
Area: 350m²
Main materials: wood, stone, wallpaper, bamboo

设计单位：江苏南通行于天装潢设计工程有限公司、石子出品高端设计事务所
设计师：孔魏躲、石小伟
项目地点：江苏南通
项目面积：350平方米
主要材料：木材、石材、墙纸、竹子

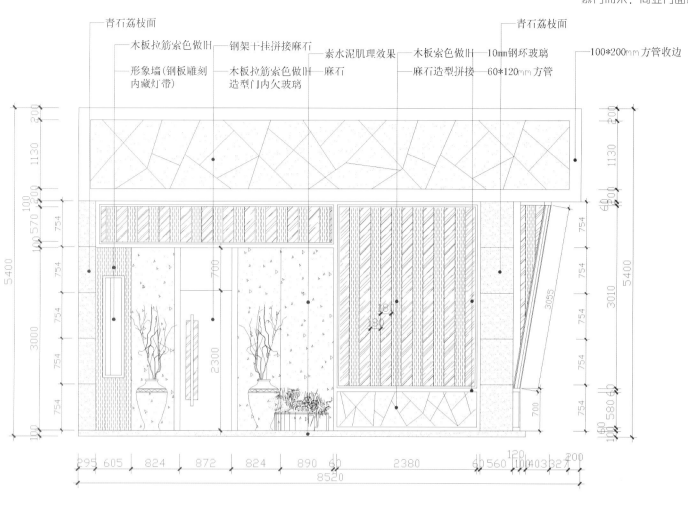

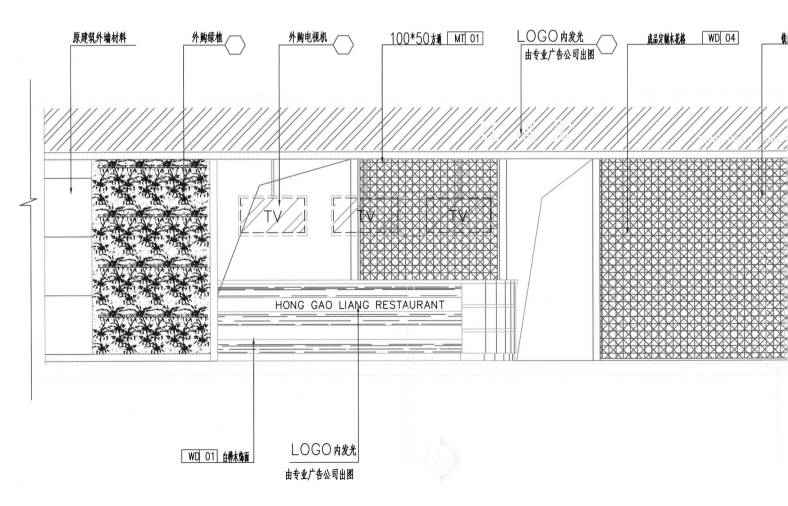

Red Sorghum Restaurant

红高粱

Design agency: Shenzhen Xiyu Design Co., Ltd.	设计单位：深圳希遇装饰设计有限公司
Designer: Zhen-hui Fu	设计师：富振辉
Location: Guangdong	项目地点：广东
Area: 540m²	项目面积：540 平方米

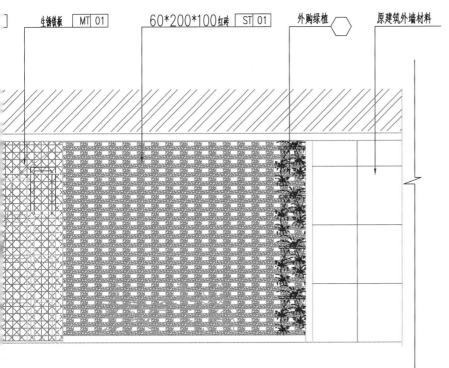

Hotpot 106

火锅 106

Design agency: Ganna Design Studio
Designer: Shih-jie Lin, Ting-liang Chen
Location: Taipei, Taiwan
Area: 125m²

设计单位：甘纳空间设计
设计师：林仕杰、陈婷亮
项目地点：台湾台北
项目面积：125 平方米

You Sushi II

游寿司二店

Design agency: AX Design Group
Location: Taipei, Taiwan

设计单位：大器联合室内装修设计有限公司
项目地点：台湾台北

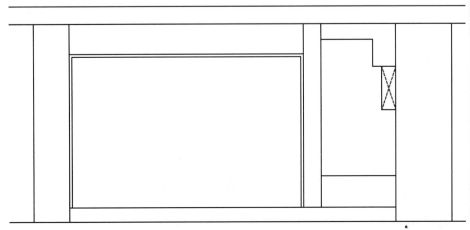

You Sushi III

游寿司三店

Design agency: AX Design Group
Location: Taipei, Taiwan

设计单位：大器联合室内装修设计有限公司
项目地点：台湾台北

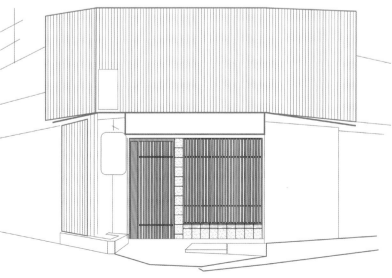

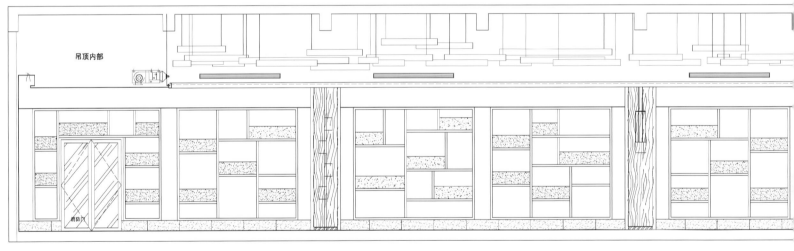

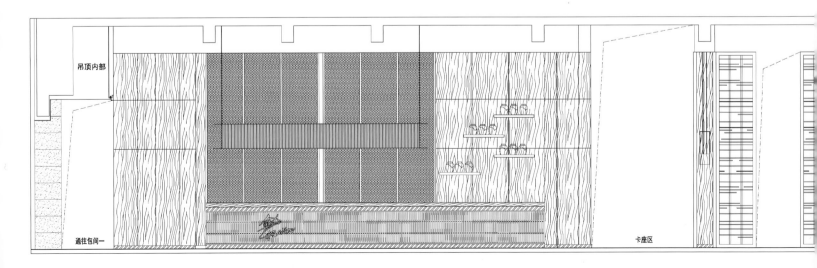

Xiang Show Restaurant

湘秀餐厅

Design agency: Yuejie Design Company
Designer: Ming Guo
Location: Beijing
Area: 700m²
Photography: Wei Yun
Main materials: ashtree, cultural stone, tile

设计单位：悦界设计
设计师：郭明
项目地点：北京
项目面积：700 平方米
摄影：恽伟
主要材料：水曲柳、文化石、地砖

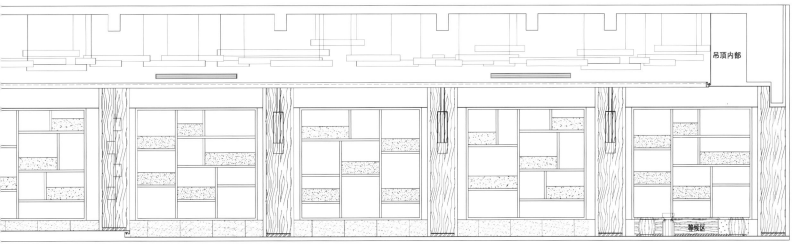

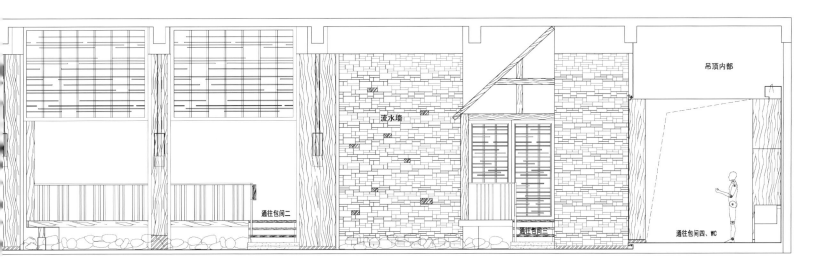

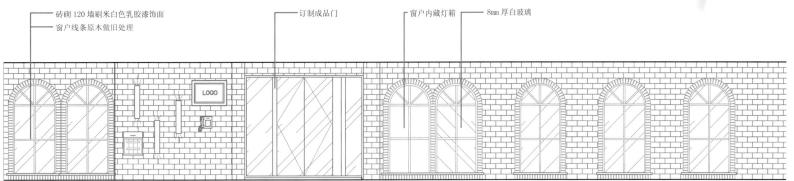

Slow Life Hot-pot
慢生活冷火锅麻辣烫

Design agency: DCV Creative Group
Designer: Yong Wang
Location: Xi'an, Shanxi
Area: 100m²
Photography: Hao Zhang
Main materials: rustic tile, laminated wooden flooring, pine board, latex paint

设计单位：DCV 第四维创意集团
设计师：王咏
项目地点：陕西西安
项目面积：100 平方米
摄影：张浩
主要材料：仿古砖、复合木地板、松木板、乳胶漆

Zhou He Teahouse
周和茗茶

Design agency: DCV Creative Group
Designer: Yong Wang
Location: Weinan, Shanxi
Area: 1000m²
Photography: Hao Zhang
Main materials: Black stone, gray mirror, gray wood tile, black cultural stone, stainless steel, wallpaper, latex paint

设计单位：DCV 第四维创意集团
设计师：王咏
项目地点：陕西渭南
项目面积：1000 平方米
摄影：张浩
主要材料：黑色石材、灰镜、灰木纹砖、黑色文化石、不锈钢、壁纸、乳胶漆

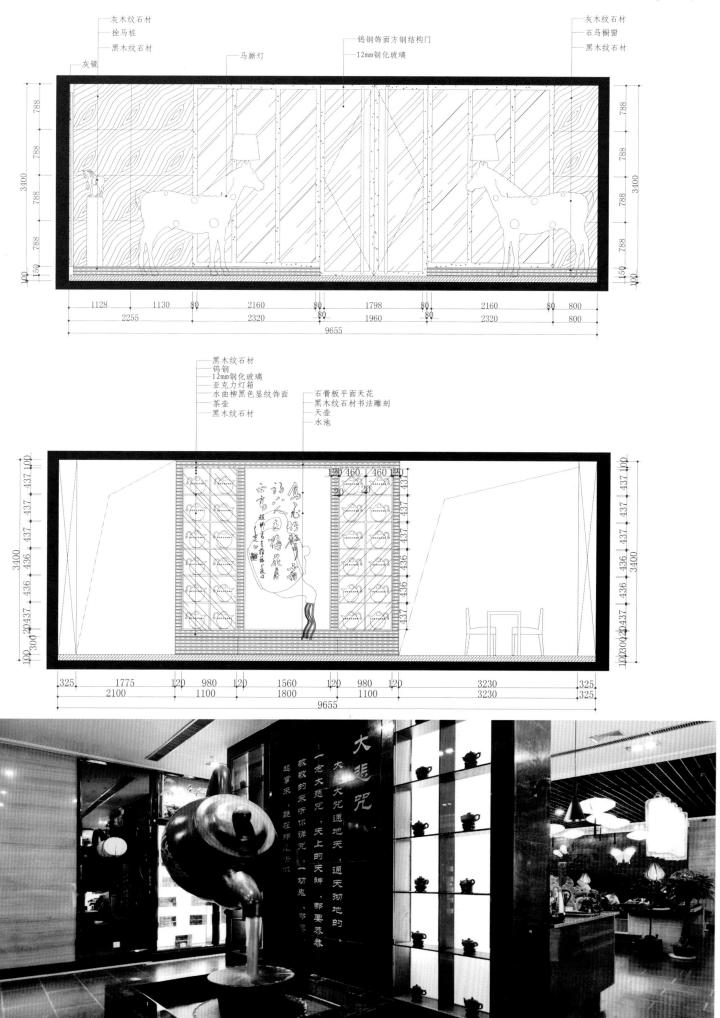

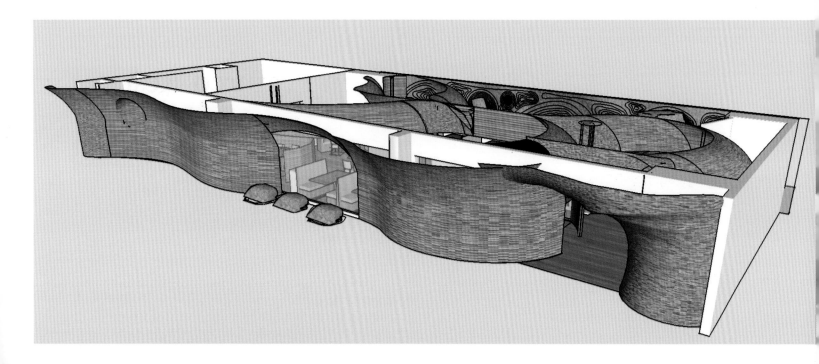

Dachu Xiaoguan Featured Restaurant

大厨小馆特色主题餐厅

Design agency: DCV Creative Group
Designer: Yong Wang, Yang-jie Hou, Tao Zhang
Location: Lijiacun, Xi' an, Shanxi
Area: 200m²
Photography: Hao Zhang, Jing-fan Duan
Main materials: PE rattan, Epoxy floor paint, rustic tile, wooden carving, colored glass, hot-dip galvanized steel

设计单位：DCV 第四维创意集团
设计师：王咏、侯洋洁、张涛
项目地点：陕西西安李家村
项目面积：200 平方米
摄影：张浩、段警凡
主要材料：PE 藤条、环氧地坪漆、仿古地砖、实木雕花、彩色玻璃、热镀锌钢骨架

Yoshinoya Fast Food Restaurant
吉野家快餐店

Design agency: AS Design Service Limited
Designer: Four Lau, Sam Sum
Location: Kowloon, Hong Kong
Area: 453m²
Photography: Sing Studio by Sum Sing

设计单位：AS Design Service Limited
设计师：刘盛科、沈浩梁
项目地点：香港九龙
项目面积：453 平方米
摄影：Sing Studio by Sum Sing

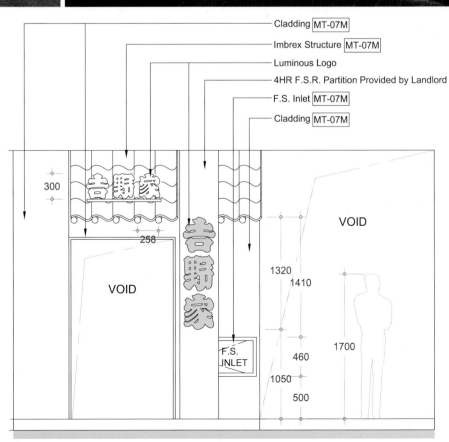

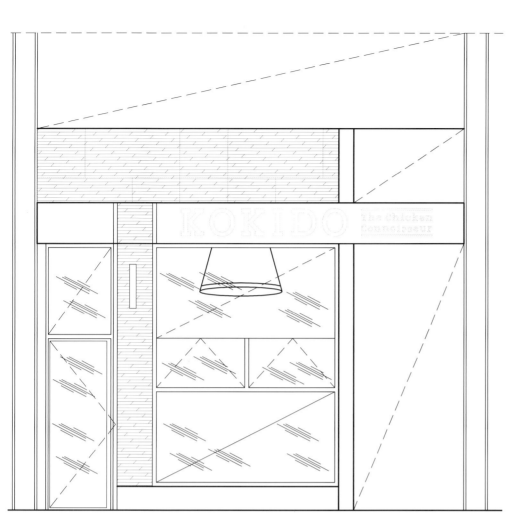

Kokido

Kokido 餐厅

Design agency: Studio Equator
Designer: Carlos Flores
Location: Caulfield, Australia
Area: 58m²
Photography: Anne-Sophie Poirier

设计单位：Equator 设计事务所
设计师：卡洛斯·弗洛雷斯
项目地点：澳大利亚考菲尔德
项目面积：58 平方米
摄影：安妮·苏菲·普瓦里耶

The Pheasantry Restaurant

The Pheasantry 餐厅

Design agency: Mizzi Studios
Location: Teddington, UK
Main materials: White Corian, Vinyl floor planks wooden effect, Siberian larch, Triple Skin Polycarbonate bespoke skylights

设计单位：米齐设计事务所
项目地点：英国特丁顿
主要材料：白色可丽耐，木纹地板革，西伯利亚落叶松，三层PC天窗

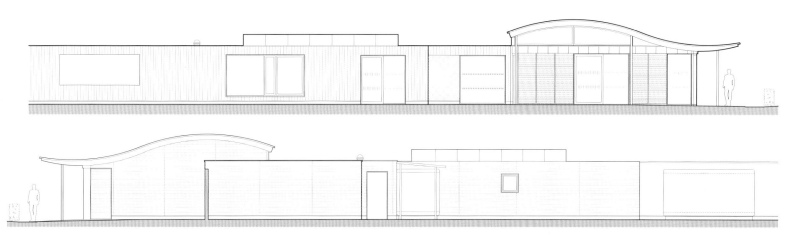

Chi Shao 100

赤烧 100

Design agency: Shenzhen Hwayon Design Consultant Limited
Designer: Zhen-zhen Huang
Location: Shenzhen, Guangdong
Area: 340m²
Main materials: red brick, laminated wooden flooring, iron artwork, wall painting, tile, cement

设计单位：深圳市华空间设计顾问有限公司

设计师：黄珍珍

项目地址：广东深圳

项目面积：340 平方米

主要材料：红砖、复合木地板、铁艺、墙绘、花砖、水泥

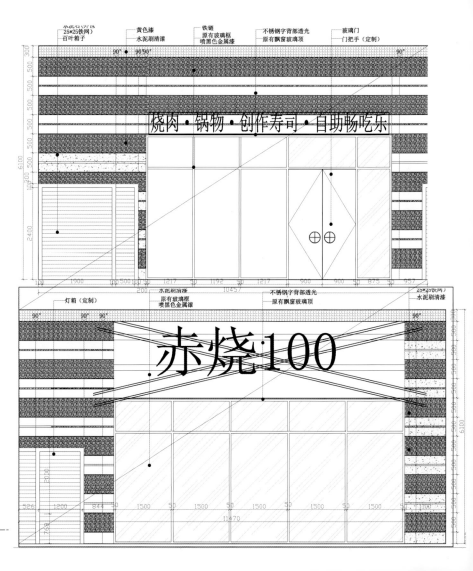

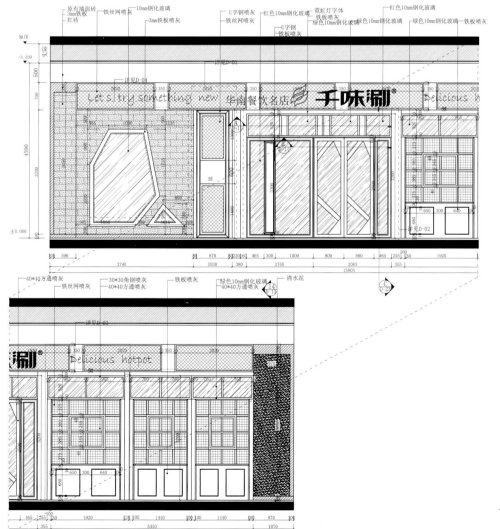

Delicious Hotpot

千味涮

Design agency: Shenzhen Hwayon Design Consultant Limited
Designer: Zheng-wei Li
Location: Chongqing
Area: 200m²
Main materials: steel, tile, concrete, wall painting, colored glass

设计单位：深圳市华空间设计顾问有限公司
设计师：李政伟
项目地址：重庆
项目面积：200平方米
主要材料：钢材、花砖、水泥、墙绘、彩色玻璃

Fish & Fusion Restaurant

Fish & Fusion 餐厅

Design agency: YOD Design Lab
Designer: Vladimir Nepiyvoda, Dmitry Bonesko
Location: Poltava, Ukraine
Area: 190 m²
Photography: Andrey Avdeenko

设计单位：YOD 设计室
设计师：Vladimir Nepiyvoda, Dmitry Bonesko
项目地点：乌克兰波尔塔瓦
项目面积：190 平方米
摄影：Andrey Avdeenko

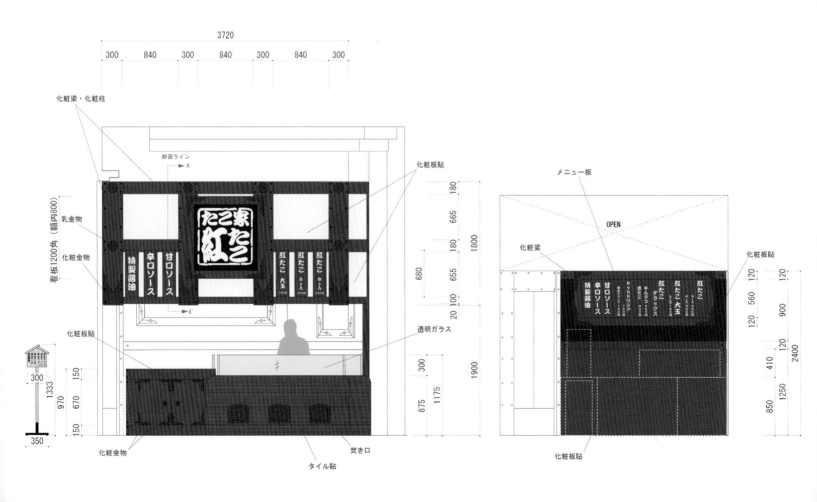

Beni Tako

红家章鱼烧

Design agency: Matsuya Art Works. Co., Ltd.	设计单位：松屋艺术设计有限公司
Designer: Tetsuya Matsumoto	设计师：Tetsuya Matsumoto
Location: Okayama, Japan	项目地点：日本冈山
Photography: Tomoki Otagakii	摄影：Tomoki Otagaki

Mr. Jiaozi

饺子先生

Designer: Jia-wei Shen
Area: 130m²

设计师：沈嘉伟
项目面积：130 平方米

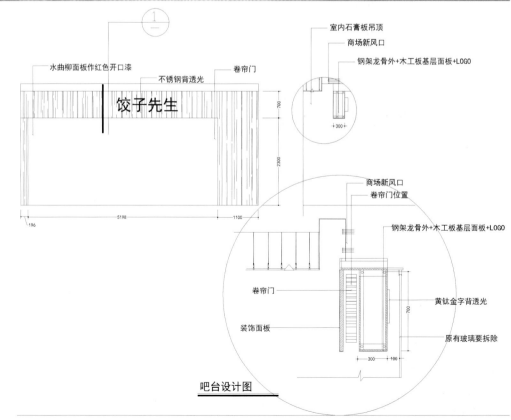

吧台设计图

Thai Cuisine

茶马天堂

Design agency: Sense Town Creative Design Co., Ltd.
Designer: Welkin Zhu
Location: Suzhou, Jiangsu

设计单位：善水堂设计
设计师：朱伟
项目地点：江苏苏州

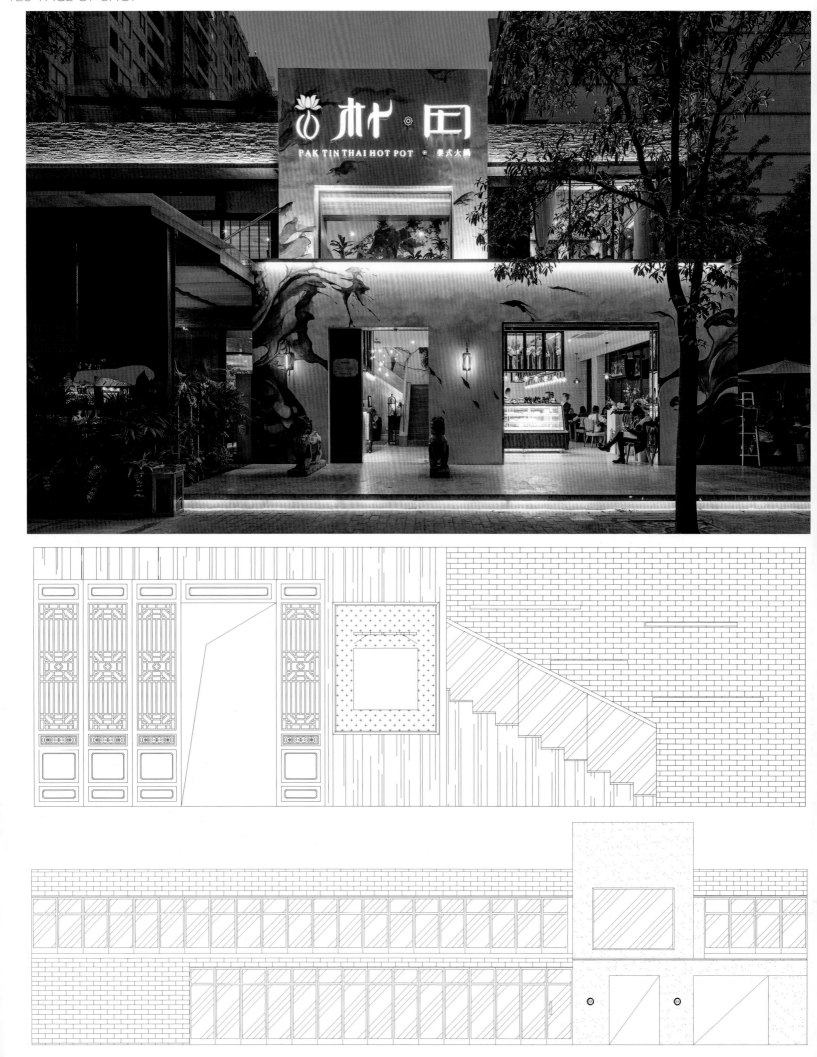

Pak Tin Thai Hot Pot
朴田泰式海鲜火锅

Designer: Jia-wei Shen
Area: 1000m²
Main materials: latex paint, cloth, white strip wall tile, stone, wallpaper, woodwork, wood flooring

设计师：沈嘉伟

项目面积：1000平方米

主要材料：乳胶漆、布艺、白色墙面条砖、荒料石材、墙纸、木作、木地板

Chosen Bun

Chosen Bun 餐厅

Design agency: Cinimod Studio
Location: London, UK
Photography: Dominic Harris

设计单位：Cinimod 设计事务所
项目地点：英国伦敦
摄影：多米尼克·哈里斯

Kessalao

Kessalao 餐厅

Design agency: Masquespacio
Location: Alemania, Germany
Area: 40m²
Photography: David Rodríguez, Carlos Huecas

设计单位：Masquespacio 设计工作室
项目地点：德国波恩
项目面积：40 平方米
摄影：大卫·罗德里格斯、卡洛斯·维卡斯

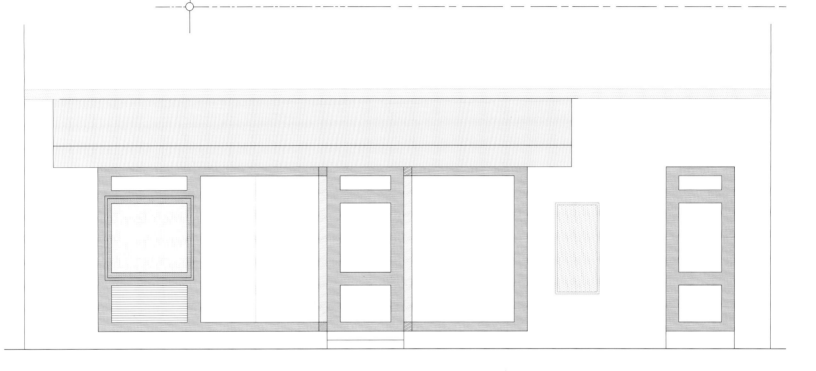

S-Michelle French Restaurant

圣米歇尔法式餐厅

Design agency: Shenzhen Disemmy Design Studio	设计单位：深圳市迪森艾美设计顾问有限公司
Designer: Sen Mu	设计师：穆森
Location: Shenzhen, Guangdong	项目地点：广东深圳
Area: 640m²	项目面积：640 平方米

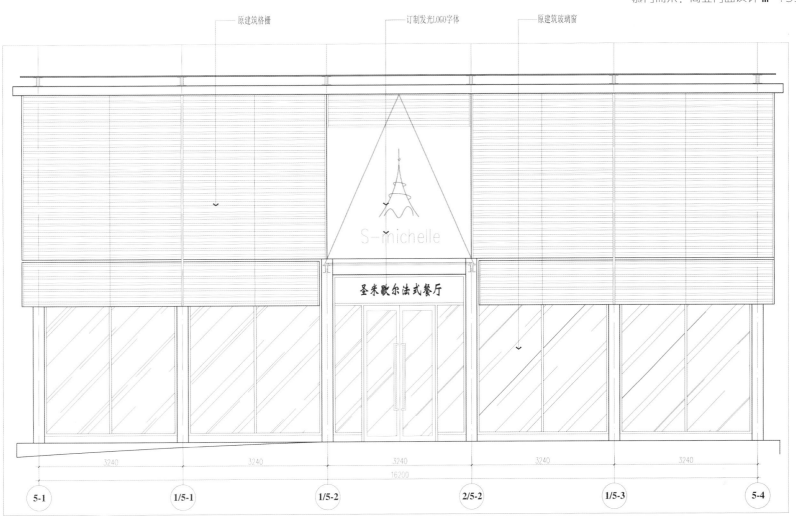

Si Cheng Xiang Shi Hot Pot

驷骋相识火锅店

Designer: Lu Lu
Location: Zhoushan, Zhejiang
Area: 500m²
Main materials: rustic tile, concrete, craft painting board, wooden shelf

设计师：卢路
项目地点：浙江舟山
项目面积：500 平方米
主要材料：复古砖、素水泥、工艺油漆木板、木架

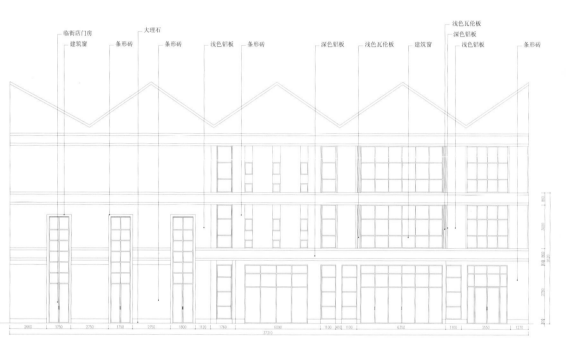

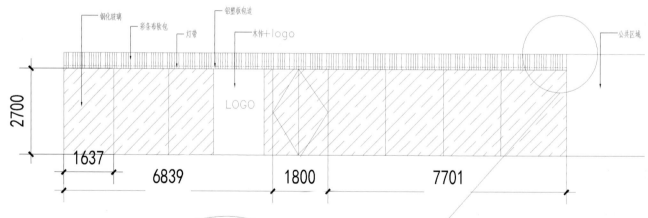

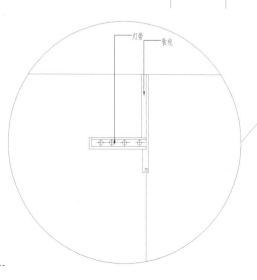

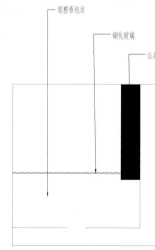

Travel Taste Buds
锦亦味餐厅

Designer: Jia-wei Shen
Location: Chengdu, Sichuan
Area: 389m²

设计师：沈嘉伟
项目地点：四川成都
项目面积：389 平方米

ZuiBaShu Sichuan Cuisine Restaurant

醉巴蜀新派川菜餐厅

Designer: Shang Wu	设计师：吴尚
Location: Dongguan, Guangdong	项目地点：广东东莞
Area: 400m²	项目面积：400 平方米
Main materials: glass, stainless steel, blue stone, wood, marble, wallpaper	主要材料：玻璃、不锈钢、青石板、木、大理石、墙纸

Grilled Fish Restaurant

有鱼吃餐厅

Designer: Jia-wei Shen
Location: Chengdu, Sichuan
Area: 130m²
Main materials: wooden flooring, wallpaper, latex paint, tile, veneer, painted glass

设计师：沈嘉伟
项目地点：四川成都
项目面积：130 平方米
主要材料：木地板、草编墙纸、乳胶漆、面砖、饰面板、喷绘玻璃

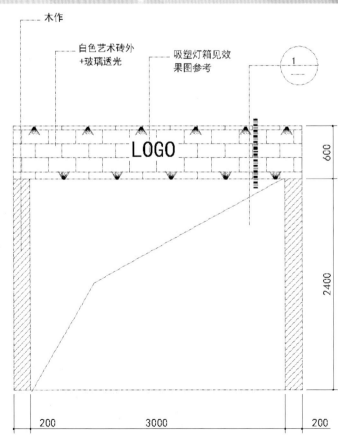

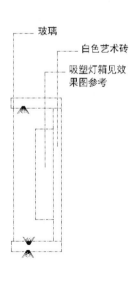

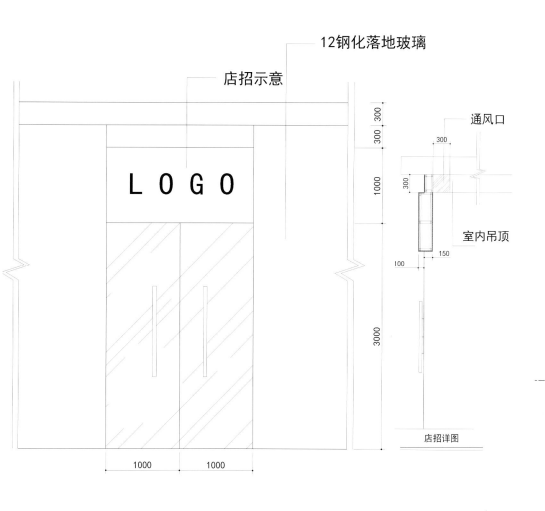

Saiho Plaza
Mr. Jiaozi
饺子先生世豪广场店

Designer: Jia-wei Shen
Location: Chengdu, Sichuan
Main materials: wooden veneer, wood flooring, wood, soft rolling, carved wood

设计师： 沈嘉伟
项目地点： 四川成都
主要材料： 木面板、实木地板、木材、软包、雕花木

Luck Sushi Bar

LUCK 寿司料理店

Designer: Jian Zhang
Location: Dalian, Liaoning
Main materials: mental mosaic, quartzite, cultural stone, mirror glass

设计师：张健
项目地点：辽宁大连
主要材料：金属马赛克、石英石、文化石、镜面玻璃

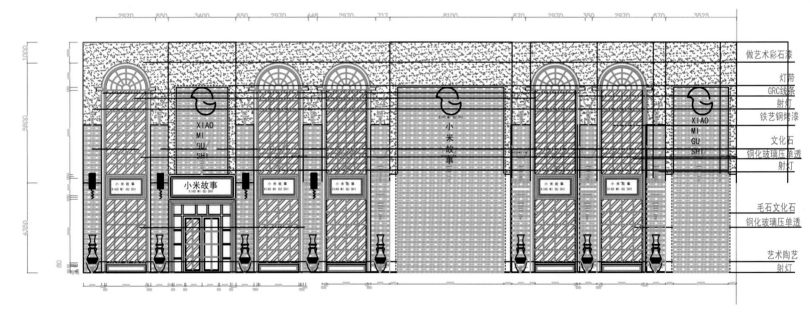

Xiaomi Gushi Restaurant

小米故事餐厅

Design agency: HK ShengJiaYou Interior Design Co., Ltd.
Designer: Dolphin Sheng
Location: Yiwu, Zhejiang
Area: 1350m²
Photography: Bighead

设计单位：香港绳家友国际设计机构
设计师：绳家友
项目地点：浙江义乌
项目面积：1350 平方米
摄影：诸华松

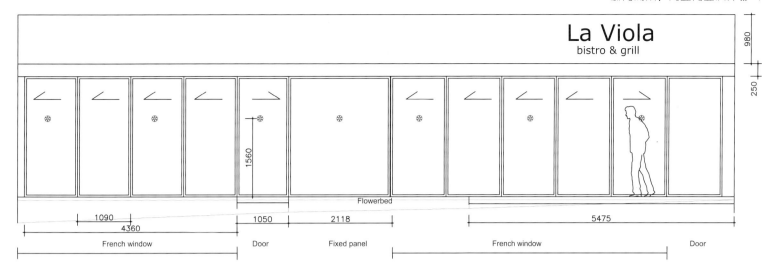

La Viola Bistro & Grill

La Viola 餐厅

Design agency: ARBOIT ltd
Designer: Alberto Puchetti
Location: Hong Kong
Area: 180m²
Photography: Dennis Lo
Main materials: concrete , woodwork , wallpaper , stainless steel , ceramic tile

设计单位：艾伯特设计有限公司
设计师：艾伯特·普切堤
项目地点：香港
项目面积：180 平方米
摄影：Dennis Lo
主要材料：混凝土、木作、定制墙纸、不锈钢、陶瓷砖

Star Burger Burger-bar

汉堡之星汉堡店

Design agency: YOD Design Lab
Designer: Vladimir Nepiyvoda, Dmitry Bonesko
Location: Kiev, Ukraine
Area: 240m²
Photography: Igor Karpenko

设计单位：YOD 设计室
设计师：Vladimir Nepiyvoda, Dmitry Bonesko
项目地点：乌克兰基辅
项目面积：240 平方米
摄影：Igor Karpenko

Kind Porridge
一品粥道

Design agency: Daohe Design
Designer: Xiong Gao
Location: Chengdu, Sichuan
Photography: Kai Shi, Ling-yu Li
Main materials: aggrandizement wood floor, white painted glass, hand-painted wall painting

设计单位：道和设计
设计师：高雄
项目地点：四川成都
摄影：施凯、李玲玉
主要材料：金刚板、白色烤漆玻璃、手绘墙画

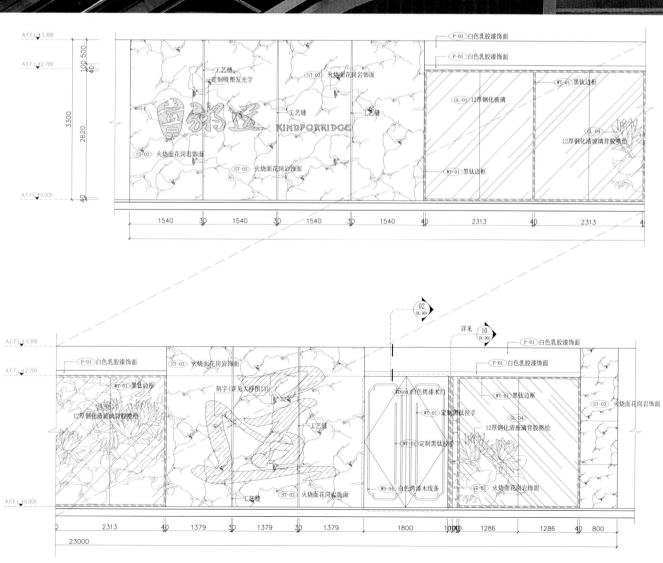

Shi Cai Restaurant
食彩餐厅

Design agency: Daohe Design
Designer: Xiong Gao
Location: Chengdu, Sichuan
Photography: Kai Shi, Ling-yu Li
Main materials: emulsification glass, painted glasses, rose-golden metal, oak wood veneer, ariston marble

设计单位：道和设计
设计师：高雄
项目地点：四川成都
摄影：施凯、李玲玉
主要材料：乳化玻璃、烤漆玻璃、玫瑰金、橡木饰面板、雅士白大理石

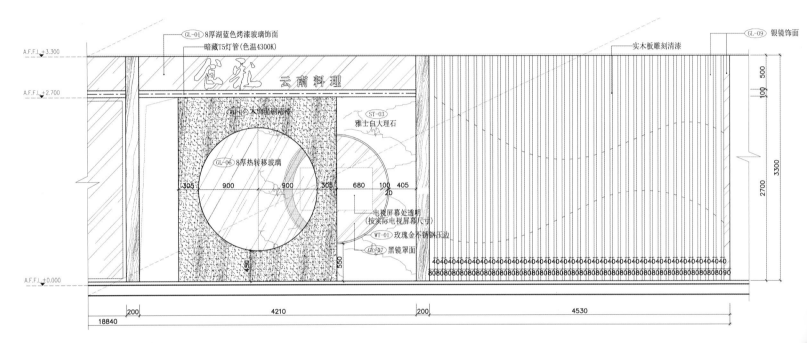

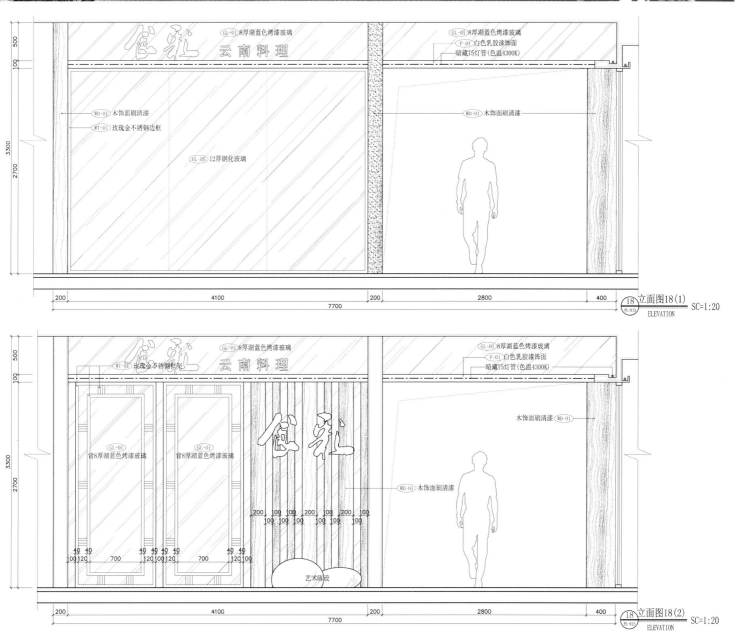

Wei Yu Restauran

味语炆泥鳅

Design agency: Daohe Design
Designer: Xiong Gao
Location: Chengdu, Sichuan

设计单位：道和设计
设计师：高雄
项目地点：四川成都

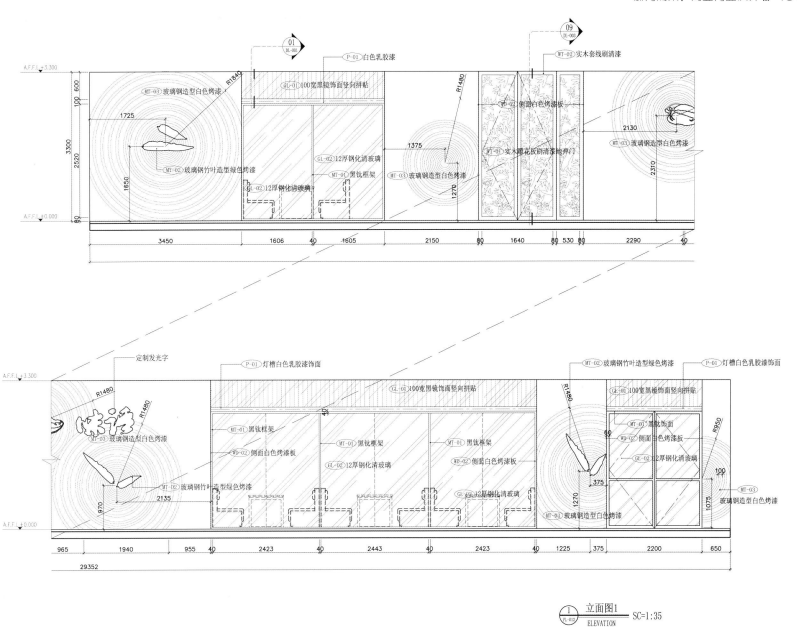

立面图1 SC=1:35
ELEVATION

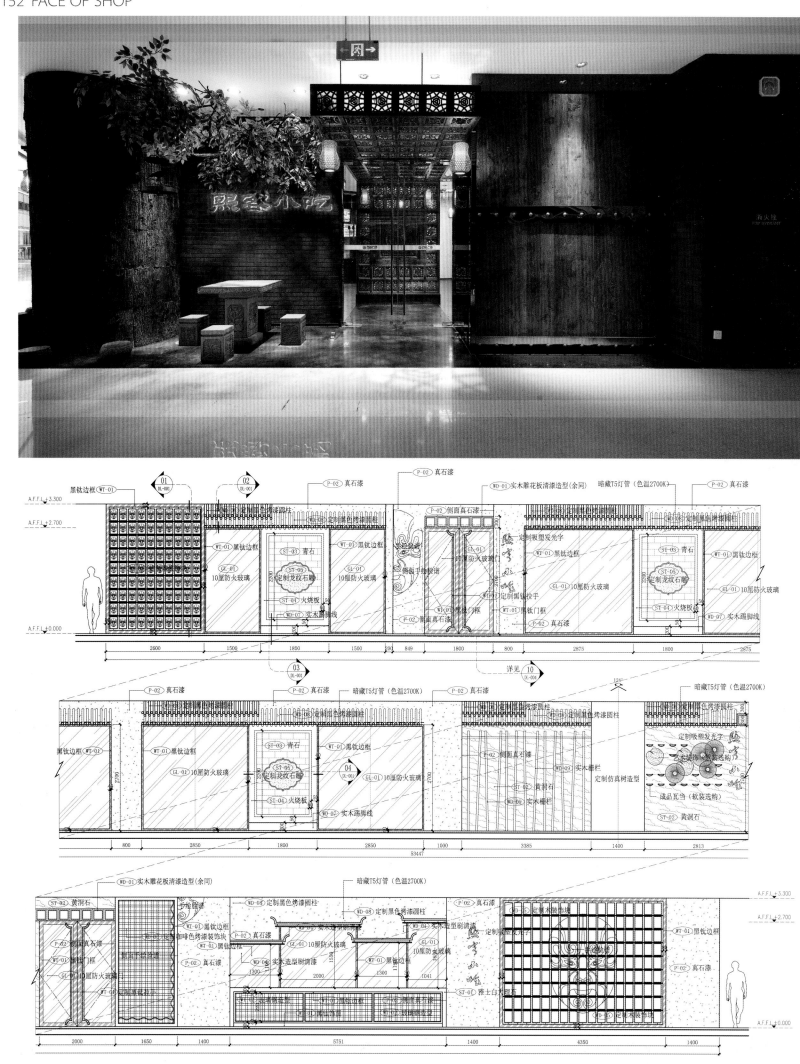

Xike Xiaochi Restaurant

熙客小吃

Design agency: Daohe Design
Designer: Xiong Gao
Location: Chengdu, Sichuan
Photography: Kai Shi, Ling-yu Li
Main materials: rustic façade tile, Volakas marble, custom wooden carved panel, cream-colored diatom ooze

设计单位：道和设计
设计师：高雄
项目地点：四川成都
摄影：施凯、李玲玉
主要材料：复古外墙砖、爵士白大理石、定制实木雕花板、米黄色硅藻泥

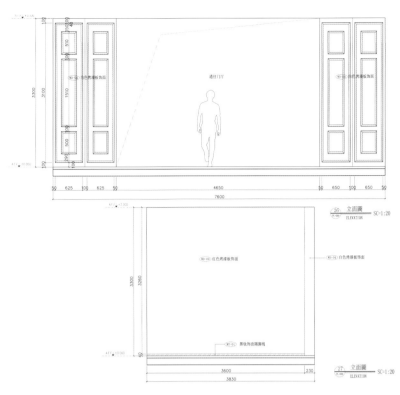

Xiang Cui Restaurant

香翠港式茶餐厅

Design agency: Daohe Design	设计单位：道和设计
Designer: Xiong Gao	设计师：高雄
Location: Chengdu, Sichuan	项目地点：四川成都
Area: 200m²	项目面积：200 平方米

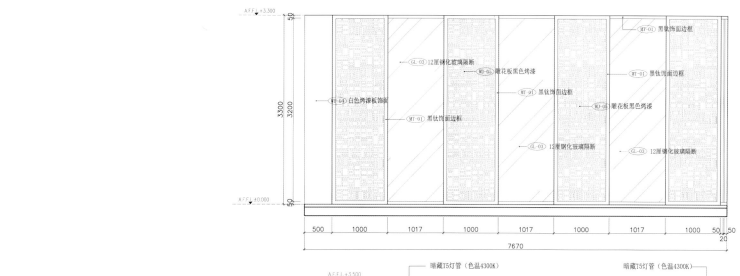

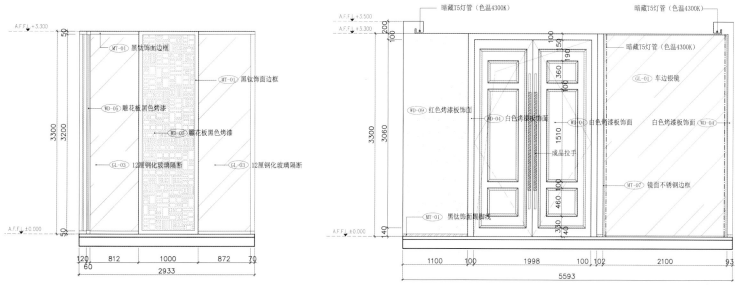

Wang Jiangnan Restaurant

望江南餐厅

Design agency: Daohe Design
Designer: Xiong Gao, Yun-zong Wu
Location: Fujian
Photography: Yue-dong Zhou

设计单位：道和设计
设计师：高雄、吴运棕
项目地点：福建
摄影：周跃东

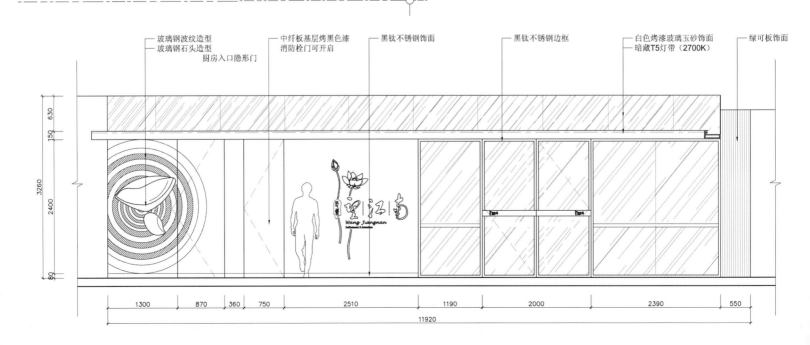

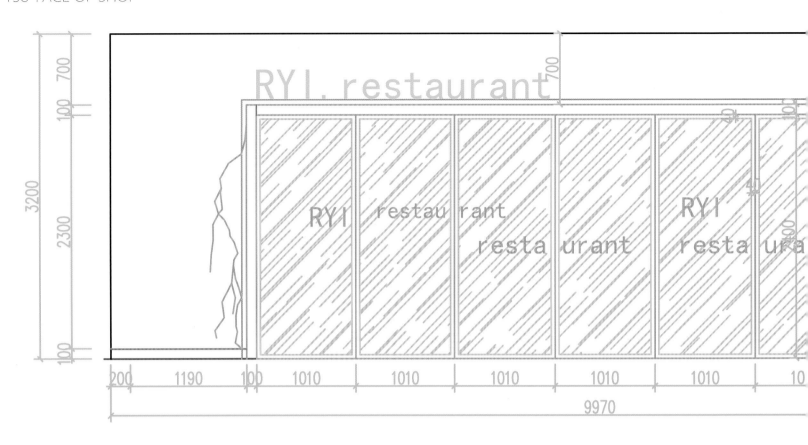

ReYi Restraunt
热意餐厅

Design agency: Trenchant (Hangzhou) Design Co., Ltd.　设计单位：杭州宣驰装饰设计有限公司
Designer: Qvent Xu　设计师：许立强
Location: Hangzhou, Zhejiang　项目地点：浙江杭州

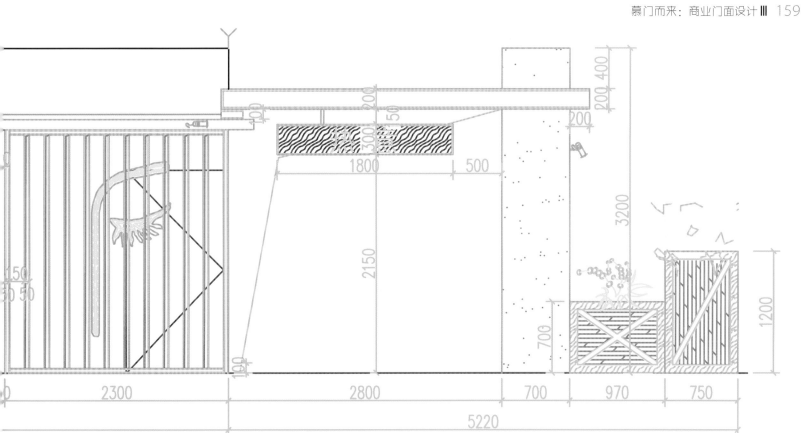

160 FACE OF SHOP

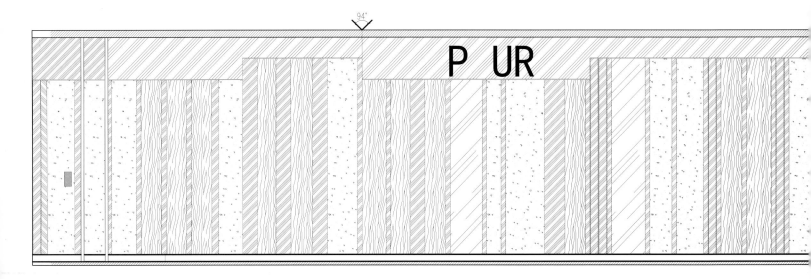

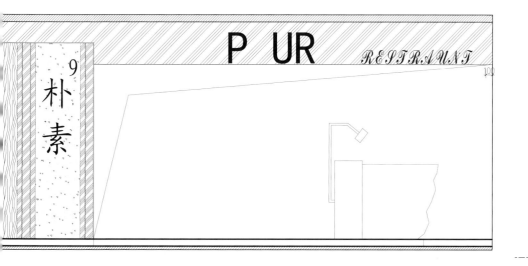

Pur Restraunt
朴素九段烧

Design agency: Trenchant (Hangzhou) Design Co., Ltd.
Designer: Qvent Xu
Location: Hangzhou, Zhejiang

设计单位：杭州宣驰装饰设计有限公司
设计师：许立强
项目地点：浙江杭州

Castle Bar
城堡餐吧

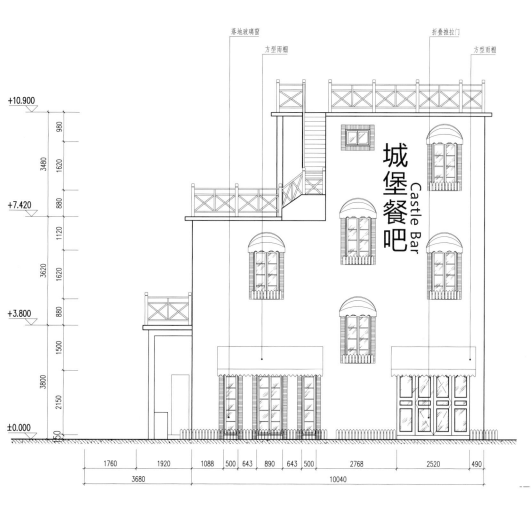

Design agency: Hong Kong Ruihe Design Co., Ltd.
Designer: Shang Wu
Location: Dongguan, Guangdong
Area: 380m²
Photography: Cheng Liu
Main materials: cultural stone, hand-made tile, preservative-treated timber, marble, veneer, colored glass

设计单位：香港瑞和装饰设计有限公司
设计师：吴尚
项目地点：广东东莞
项目面积：380 平方米
摄影：刘诚
主要材料：文化石、手工砖、防腐木、大理石、饰面板、彩色玻璃

Zoe Restaurant

尚渝餐厅

Design agency: IGIG Creative
Designer: Ran Xu, Hui-ying Qu
Location: Shanghai
Area: 300m²
Main materials: handmade ceramic plate, gray serpeggianto, imported resin plate

设计单位：旋木空间设计
设计师：冉旭、屈慧颖
项目地点：上海
项目面积：300平方米
主要材料：手工陶板、灰色木纹石、进口树脂板

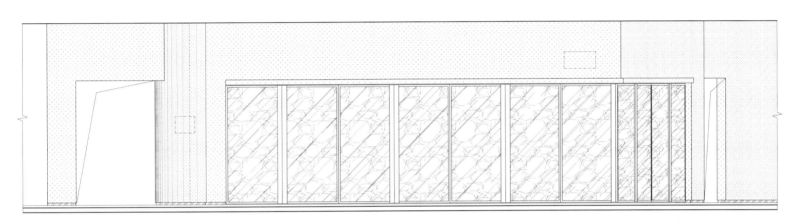

Great Food Hall
Great 美食广场

Design agency: HEAD Architecture and Design
Location: Hong Kong

设计单位：HEAD 建筑设计有限公司
项目地点：香港

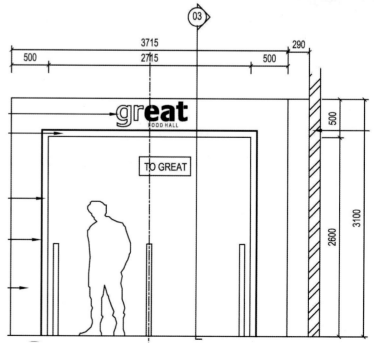

HaiFuGang Japanese Cuisine

海福冈日本料理店

Design agency: Home Young Interior Design
Designer: Hui-hsin Cheng
Location: Taipei, Taiwan
Area: 50m²

设计单位：鸿样空间有限公司
设计师：郑惠心
项目地点：台湾台北
项目面积：50平方米

ChanShi Restaurant

禅石餐厅

Design agency: Wuhan Pinzhu LingChuan Design Consulting Co., Ltd.
Designer: Chuan Ling
Location: Wuhan, Hubei
Area: 912m²
Photography: Zhi-bing Yuan

设计单位：武汉品筑凌川设计顾问有限公司
设计师：凌川
项目地点：湖北武汉
项目面积：912 平方米
摄影：袁知兵

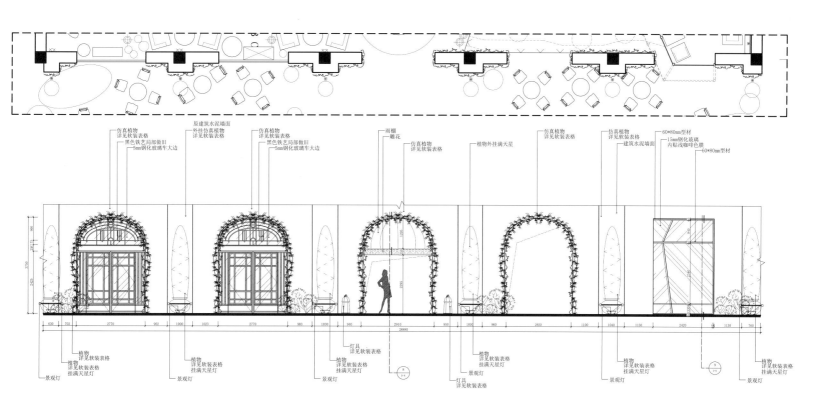

Abeille Cake

Abeille 蛋糕店

Design agency: Matsuya Art Works. Co., Ltd.	设计单位：松屋艺术设计有限公司
Designer: Tetsuya Matsumoto	设计师：Tetsuya Matsumoto
Location: Himeji, Japan	项目地点：日本姬路市
Photography: Toshiyuki Nishimatsu	摄影：Toshiyuki Nishimatsu

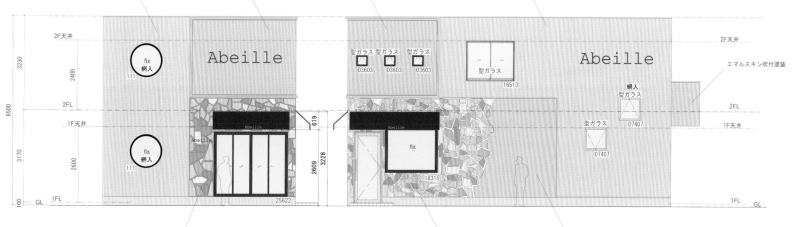

Hisago Omusubi

Hisago 饭团铺

Design agency: Matsuya Art Works. Co., Ltd.
Designer: Tetsuya Matsumoto
Location: Himeji, Japan
Photography: Toshiyuki Nishimatsu

设计单位：松屋艺术设计有限公司
设计师：Tetsuya Matsumoto
项目地点：日本姬路市
摄影：Toshiyuki Nishimatsu

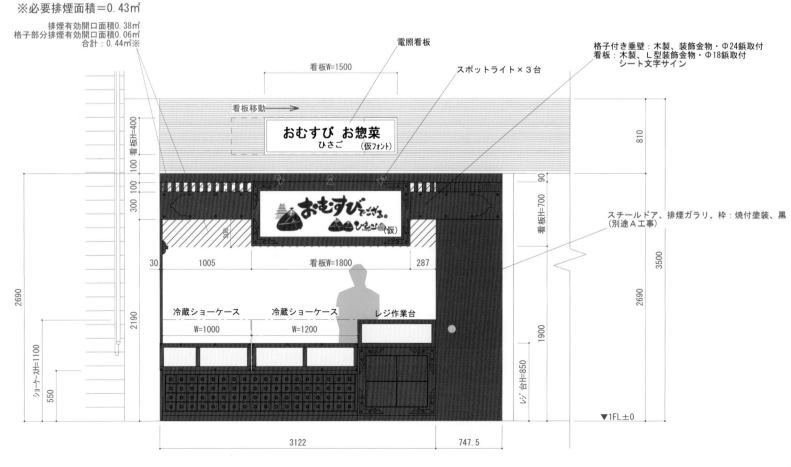

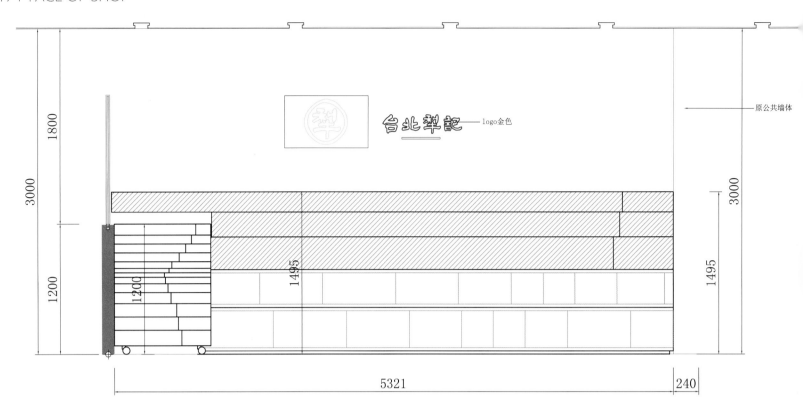

Taipei Lee Chi
台北犁记

Design agency: 1/10 Concept Programming Interior
Designer: Tracy Jen
Location: Shanghai
Area: 16m²
Photography: Zhen-yu Lu

设计单位：十分之一设计事业有限公司
设计师：任萃
项目地点：上海
项目面积：16 平方米
摄影：卢震宇

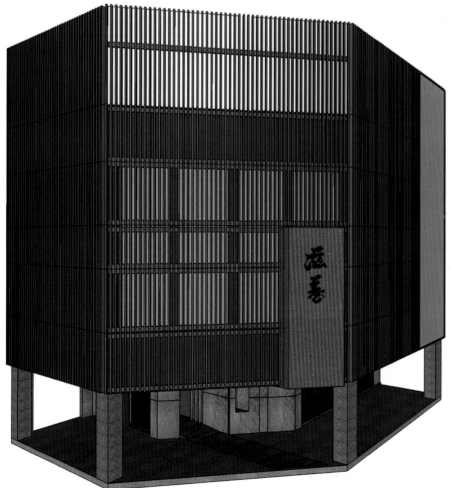

Ziyang Japanese Desert Store

滋养和果子店

Design agency: AX Design Group
Location: Taipei, Taiwan
Area: 162m²

设计单位：大器联合室内装修设计有限公司
项目地点：台湾台北
项目面积：162平方米

水文化：咖啡、茶、酒
Water Culture: Coffee, Tea, Wine

Annvita Tea House
安薇塔茶屋

Design agency: Shanghai Scale Art Design Corporation
Designer: Jian-liang Lai
Location: Beijing
Area: 1200m²

设计单位：上海十方圆国际设计工程
设计师：赖建良
项目地点：北京
项目面积：1200 平方米

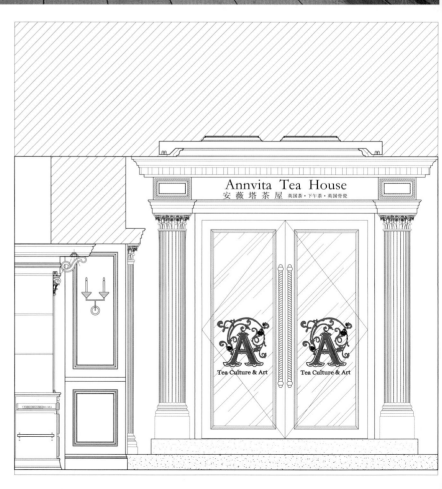

Cafe Football
足球咖啡厅

Design agency: Checkland Kindleysides	设计单位：Checkland Kindleysides 设计顾问公司
Location: London, UK	项目地点：英国伦敦

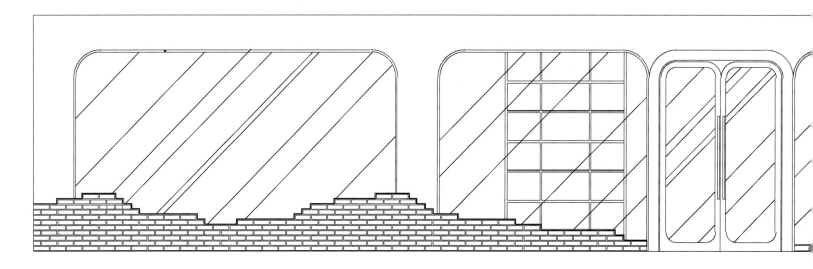

C'est la Vie café
乐菲咖啡

Design agency: Jiangsu Nantong XingYuTian Design & Decoration Engineering Co., Ltd.; ShiZi ChuPin Design Studio
Designer: Wei-duo Kong, Xiao-wei Shi
Location: Nantong, Jiangsu
Area: 255m²
Main materials: wood, complex stone veneer, concrete

设计单位：江苏南通行于天装潢设计工程有限公司、石子出品高端设计事务所
设计师：孔魏躲、石小伟
项目地点：江苏南通
项目面积：255 平方米
主要材料：木材、复合石材、素水泥

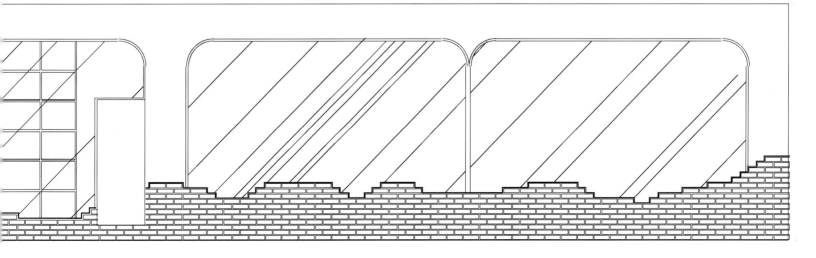

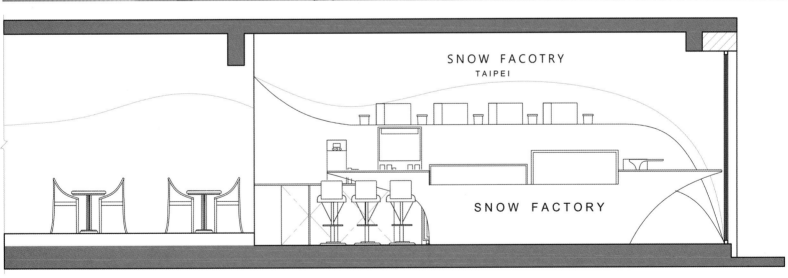

Snow Factory

雪坊优格

Design agency: 1/10 Concept Programming Interior
Designer: Tracy Jen
Location: Taipei, Taiwan

设计单位：十分之一设计事业有限公司
设计师：任萃
项目地点：台湾台北

N.O.T. Specialty Coffee

N.O.T. Specialty 咖啡店

Design agency: Atelier E Limited
Designer: Nuo Xu
Location: Hong Kong
Area: 20m²

设计单位：Atelier E Limited 设计事务所
设计师：许诺
项目地点：香港
项目面积：20 平方米

Hazel & Hershey Cafe

Hazel & Hershey 咖啡厅

Design agency: Atelier E Limited
Designer: Nuo Xu
Location: Hong Kong
Area: 100m²

设计单位：Atelier E Limited 设计事务所
设计师：许诺
项目地点：香港
项目面积：100 平方米

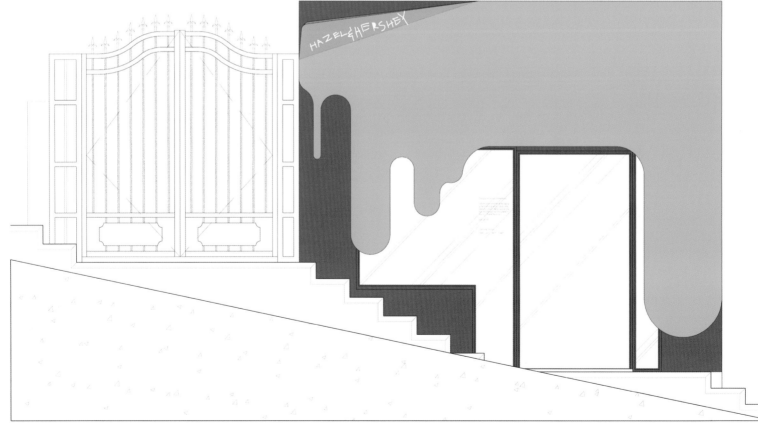

Wu Yu Tai Tea Salon

吴裕泰茶荟观

Design agency: Shanghai Scale Art Design Corporation
Designer: Jian-liang Lai
Location: Shanghai
Area: 475m²

设计单位：上海十方圆国际设计工程
设计师：赖建良
项目地点：上海
项目面积：475 平方米

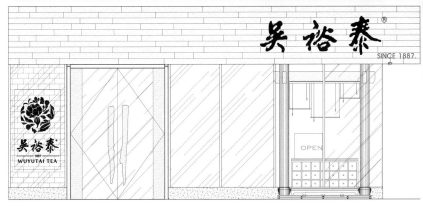

Opus Cafe
作品咖啡

Design agency: Sense Town Creative Design Institution
Location: Suzhou, Jiangsu

设计师：善水堂创意设计机构
项目地点：江苏苏州

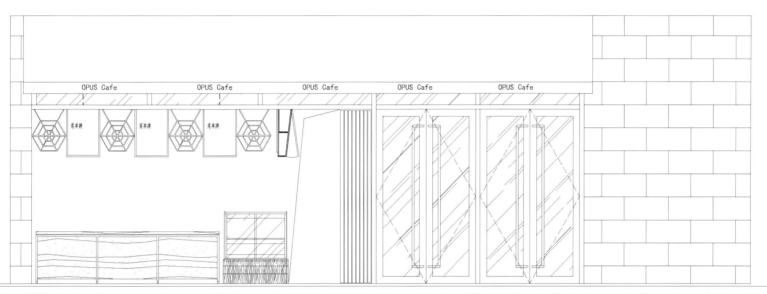

Milk Tea & Pearl Boxpark
Boxpark 奶茶店

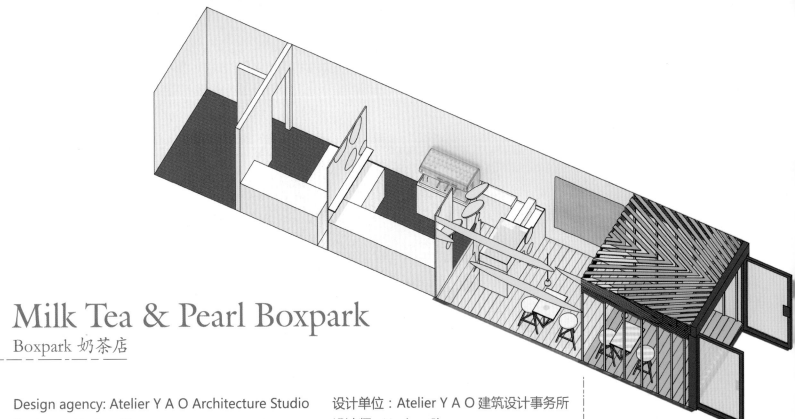

Design agency: Atelier Y A O Architecture Studio
Designer: Yaojen Chuang
Location: London, UK
Area: 28m²
Photography: Atelier Y A O

设计单位：Atelier Y A O 建筑设计事务所
设计师：Yaojen Chuang
项目地点：英国伦敦
项目面积：28 平方米
摄影：Atelier Y A O

Mocoway Coffee House
Mocoway 咖啡馆

Design agency: ONG&ONG Pte Ltd.
Designer: David Huang, Kenny Liu
Location: Chengdu, Sichuan
Photography: Lang Sha Photo & Design Studio

设计单位：王及王建筑事务所
设计师：David Huang, Kenny Liu
项目地点：四川成都
摄影：朗萨映画摄影设计工作室

Time Art Coffee
时艺汇咖啡厅

Design agency: Wang Feng-bo Design Studio	设计单位：北京王凤波装饰设计机构出品
Designer: Feng-bo Wang	设计师：王凤波
Location: Beijing	项目地点：北京
Area: 260m²	项目面积：260 平方米

Mansion Coffee

慢象咖啡

Design agency: Daohe Design
Designer: Xiong Gao
Location: Fujian

设计单位：道和设计
设计师：高雄
项目地点：福建

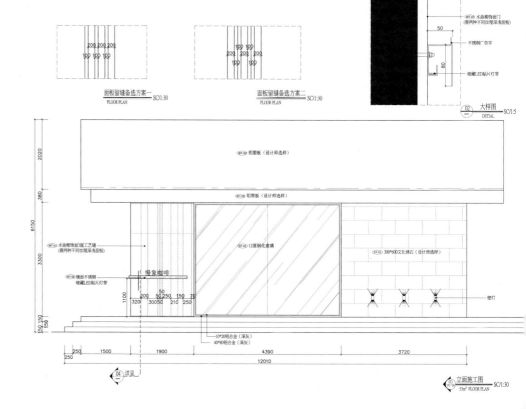

Snog Pure Frozen Yogurt Bus

SNOG 酸奶冰淇淋巴士

Design agency: Cinimod Studio
Location: London, UK
Photography: Kerim

设计单位：Cinimod 设计事务所
项目地点：英国伦敦
摄影：Kerim

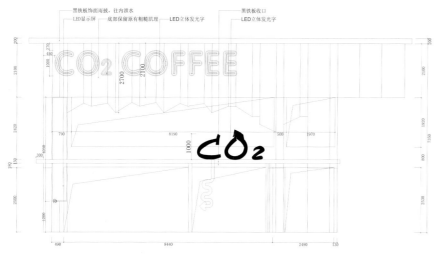

CO_2 Coffee
CO_2 咖啡厅

Design agency: Ou Di Bao Design &Decoration Co., Ltd.; Jian Jing Design Engineering Co., Ltd.
Designer: Qing-liang Chen, Feng-biao Guo
Location: Xiamen, Fujian
Area: 480m²

设计单位：欧迪堡设计装饰有限公司 / 渐境设计装饰工程有限公司
设计师：陈清凉、郭峰标
项目地点：福建厦门
项目面积：480 平方米

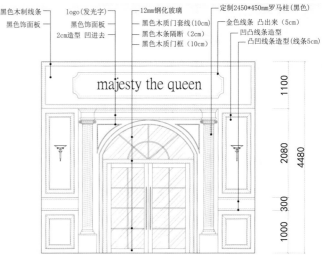

Majesty The Queen

女王陛下英式茶

Design agency: Ningbo He Gongshe Interior Design
Designer: Qin-wei Hu
Location: Ningbo, Zhejiang
Area: 108m²
Photography: Ying Liu
Main materials: wood veneer, cocrete panel, rome column, artustic floor

设计单位：宁波市禾公社装饰设计有限公司
设计师：胡秦玮
项目地点：浙江宁波
项目面积：108 平方米
摄影：刘鹰
主要材料：木墙板、水泥板、罗马柱、艺术地板

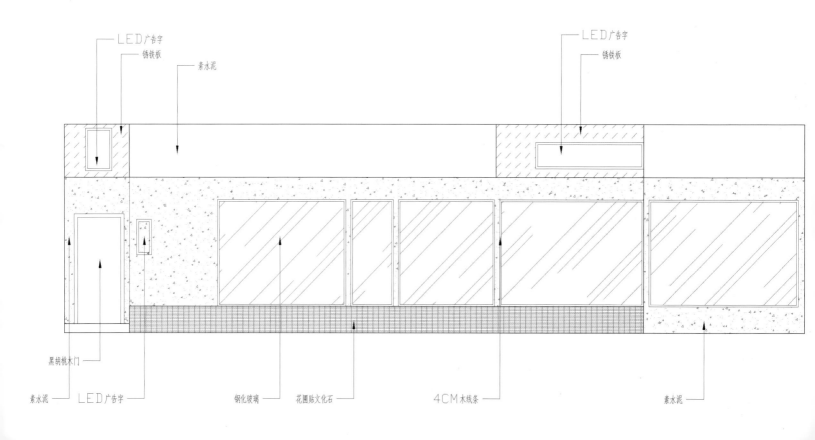

Buddhi and Tea Zen Tea House

一茶一菩提禅茶坊

Designer: Hang Yuan
Location: Ningbo, Zhejiang
Photography: Tian-hong Yu
Main materials: concrete, marble, bluestone, rosewood, pinewood, flooring coating, stainless steel

设计师：袁航
项目地点：浙江宁波
摄影：虞天鸿
主要材料：素水泥、大理石、青石板、酸枝木、老杉木、地坪漆、不锈钢

Vermillion Bar & Restaurant

弗米利恩酒吧餐厅

Design agency: Matsuya Art Works. Co., Ltd.
Designer: Tetsuya Matsumoto
Location: Himeji, Japan

设计单位：松屋艺术设计有限公司
设计师：Tetsuya Matsumoto
项目地点：日本姬路市

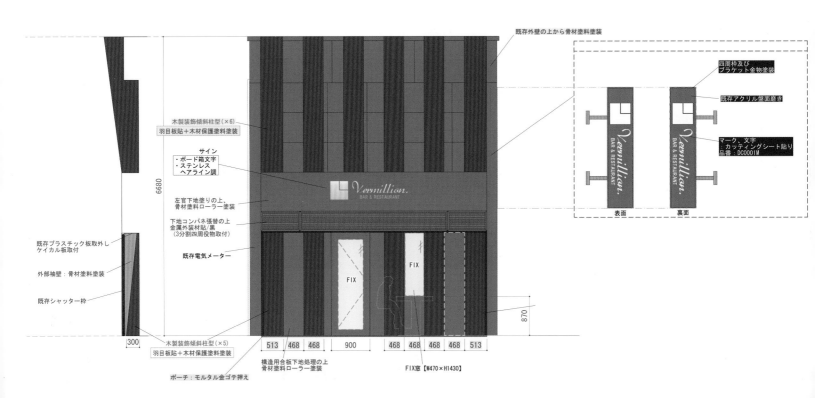

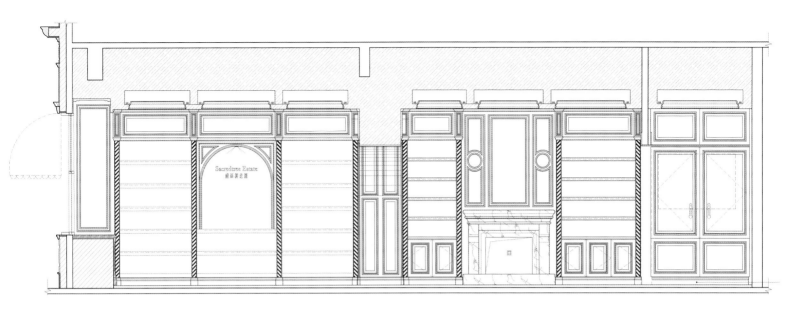

Gemtree Wine Club
宝石树红酒会所

Design agency: Dov Design Studio
Designer: Xiao-ying Zhang, Bin Fan, Qian Ao
Location: Chengdu, Sichuan
Area: 600m²

设计单位：多维设计事务所
设计师：张晓莹、范斌、敖谦
项目地点：四川成都
项目面积：600 平方米

Comfort Zone

乐酒吧

Design agency: Jiangsu Nantong XingYuTian Design & Decoration Engineering Co., Ltd.; ShiZi ChuPin Design Studio	设计单位：江苏南通行于天装潢设计工程有限公司、石子出品高端设计事务所
Designer: Wei-duo Kong, Xiao-wei Shi	设计师：孔魏躲、石小伟
Location: Nantong, Jiangsu	项目地点：江苏南通
Area: 150m²	项目面积：150 平方米
Main materials: wood, stone	主要材料：木材、石材

Monster Club
猛斯特酒吧

Design agency: Jiangsu Nantong XingYuTian Design & Decoration Engineering Co., Ltd.; ShiZi ChuPin Design Studio
Designer: Wei-duo Kong, Xiao-wei Shi
Location: Nantong, Jiangsu
Area: 300m²
Main materials: wood, stone, steel tube

设计单位：江苏南通行于天装潢设计工程有限公司、石子出品高端设计事务所
设计师：孔魏躲、石小伟
项目地点：江苏南通
项目面积：300平方米
主要材料：木材、石材、钢管

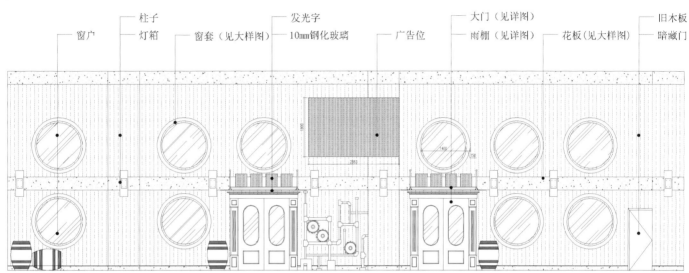

LenKa Bar
LenKa 清吧

Design agency: Yu Ze Design
Designer: Wei-min Xiao
Location: Nanjing, Jiangsu
Area: 70m²
Main materials: carbonized wood floor, anticorrosive wood wall, blue stone

设计单位：宇泽设计
设计师：肖为民
项目地点：江苏南京
项目面积：70 平方米
主要材料：炭化木地板、防腐木墙面、青石板

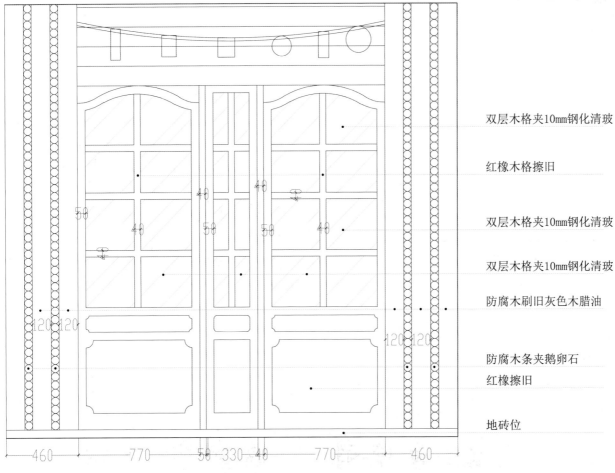

双层木格夹10mm钢化清玻

红橡木格擦旧

双层木格夹10mm钢化清玻

双层木格夹10mm钢化清玻

防腐木刷旧灰色木腊油

防腐木条夹鹅卵石

红橡擦旧

地砖位

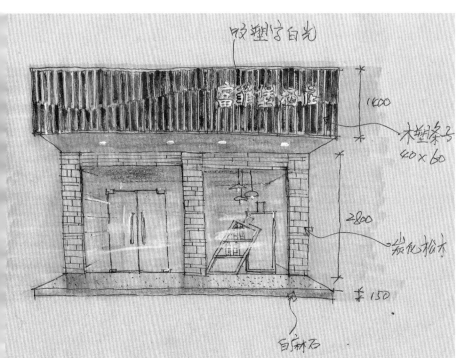

FFB Red Wine
富菲堡红酒庄

Designer: Richard Hong	设计师：康世畅
Location: Haifeng, Guangdong	项目地点：广东海丰
Photography: Wei-jin Ou	摄影：欧伟金

AMAZING CLUB Bar

AMAZING CLUB 清酒吧

Design agency: Inear Design & Decoration
Designer: Xiao-ming Zhu
Location: Hangzhou, Zhejiang
Photography: Feng Lin
Main materials: recycled wood panel, aluminium plate, steel plate, red brick, pure concrete

设计单位：杭州意内雅建筑装饰设计有限公司
设计师：朱晓鸣
项目地点：浙江杭州
摄影：林峰
主要材料：回购老木板、铝板、钢板、红砖、素水泥

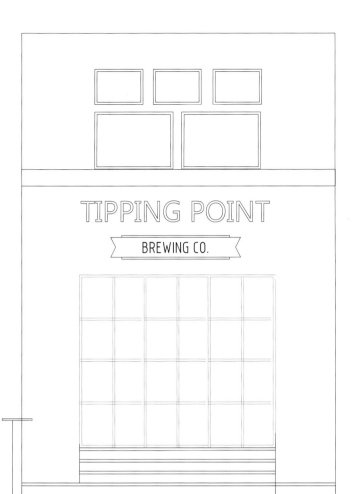

Tipping Point
Tipping Point 啤酒吧

Design agency: Arboit ltd.
Designer: Alberto Puchetti
Location: Hong Kong
Area: 120m²
Photogtaphy: Dennis Lo

设计单位：艾伯特设计有限公司
设计师：艾伯特·普切堤
项目地点：香港
项目面积：120 平方米
摄影：Dennis Lo

快乐养生：会所
Happy Regimen:Club

One Taste Holistic Health

一味坊心灵会所

Design agency: CROX International Co., Ltd.
Designer: Tsung-jen Lin, Ying-xiu Lin, Ben-tao Li
Location: Hangzhou, Zhejiang
Area: 360m²
Photography: Ji-shou Wang
Main materials: teak floor, cedarwood ceiling, philippine lauan, grain marquina floor, mirror, silver foil ceiling

设计单位：阔合国际有限公司
设计师：林琮然、林盈秀、李本涛
项目地点：浙江杭州
项目面积：360平方米
摄影：王基守
主要材料：柚木地板、杉木天花、柳安木、木纹石地面、镜子、银箔天花

Nature in all it's beauty, comes with dangers that we as humans are no longer adapted to.

Civility comfort with in our boxed off concrete settings deny us the pleasure of being apart of nature.

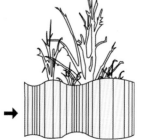

Relaxation a more fluid setting is the most comfortable enviroment for humans to be apart of nature whilst maintaining the civility that we have grown accustomed to.

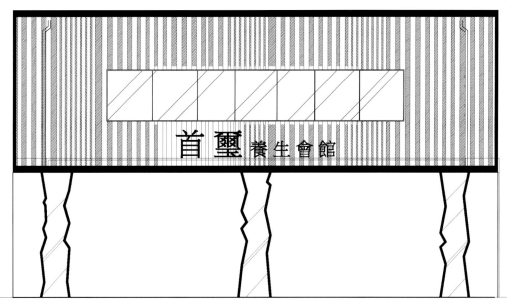

Capital Spa & Healthy Club

首玺美容养生会馆

Design agency: DAJ Interior Design Co., Ltd.	设计单位：大间空间设计有限公司
Location: Taichung, Taiwan	项目地点：台湾台中
Area: 390m²	项目面积：390 平方米
Main materials: tawny mirror, silk curtain, semi-reflective mirror glass, plastic tile, oak bark, hollow plate	主要材料：茶镜、丝帘、银半反射镜面玻璃、塑料地砖、橡木皮、中空板

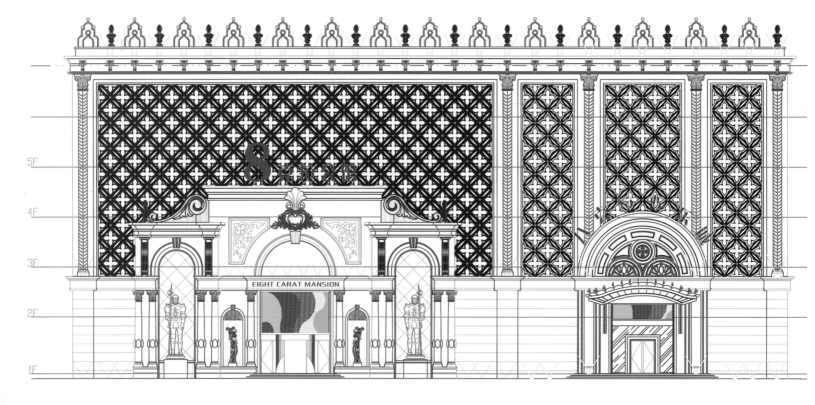

慕门而来：商业门面设计 Ⅲ 217

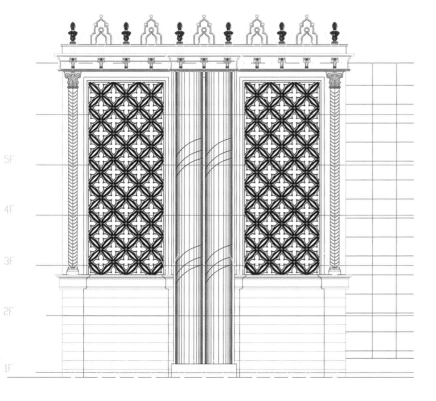

石家庄八克拉公馆

Design agency: Beijing Rhythm Space Interior Design Studio
Designer: Zhe-hao Jiang
Location: Beijing

设计单位：北京韵空间装饰设计工作室
设计师：姜哲浩
项目地点：北京

ShangShan Yoga JingShe

上善瑜伽精舍

Design agency: Jiangsu Nantong XingYuTian Design & Decoration Engineering Co., Ltd; ShiZi ChuPin Design Studio
Designer: Wei-duo Kong, Xiao-wei Shi
Location: Nantong, Jiangsu
Area: 400m²
Main materials: wood, stone, wallpaper

设计单位：江苏南通行于天装潢设计工程有限公司、
石子出品高端设计事务所
设计师：孔魏躲、石小伟
项目地点：江苏南通
项目面积：400 平方米
主要材料：木材、石材、墙纸

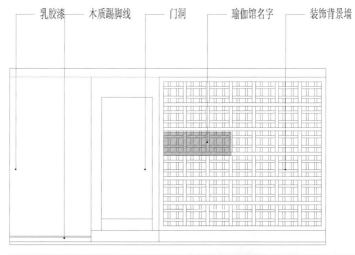

环球五号会所

Design agency: Shenzhen 0755 Decorate Design
Designer: Michael
Location: Chongqing
Area: 3500m²
Photography: Michael
Main materials: marble, ceramic tile, stainless steel, glass

设计单位：深圳市零柒伍伍装饰设计有限公司
设计师：黄治奇
项目地点：重庆
项目面积：3500 平方米
摄影：黄治奇
主要材料：大理石、瓷砖、不锈钢、玻璃

Suzhou Fanhua Jinyi Cinemas

金逸影城苏州繁花电影院

Design agency: XYI Design Consulting Co., Ltd.
Designer: Mac Huang
Location: Suzhou, Jiangsu
Area: 4750m²
Photography: Tao Wang
Main materials: Juglans cinerea, black iron, mirror, damping material, white paint, marble, glazed tile, blended carpet, black painted glass

设计单位：隐巷设计顾问有限公司
设计师：黄士华
项目地点：江苏苏州
项目面积：4750 平方米
摄影：王涛
主要材料：胡桃木染灰、黑铁、明镜、吸音材料、白色烤漆、大理石、抛光砖、混纺地毯、黑色烤漆玻璃

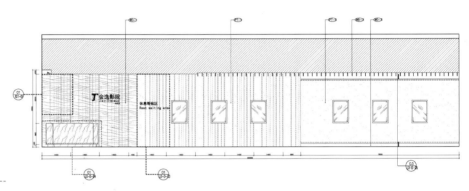

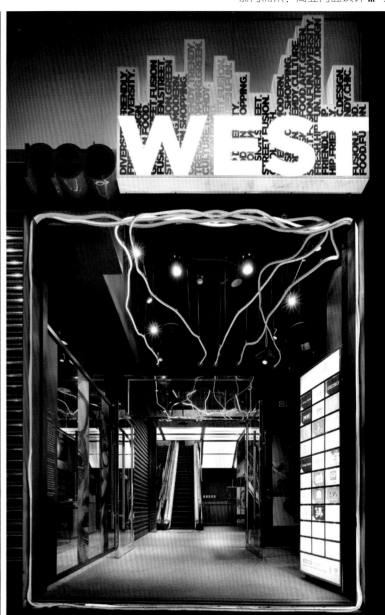

Neo West

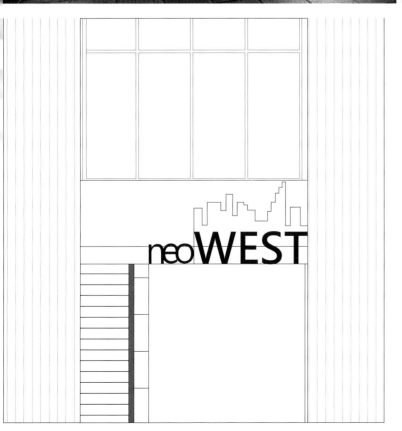

| Design agency: 1/10 Concept Programming Interior
| Designer: Tracy Jen
| Location: Taipei, Taiwan
| Area: 920m²
| Photography: Zhen-yu Lu
| Main materials: stucco washing finish, matt black painted board, mirror, tile, white artificial stone

设计单位：十分之一设计事业有限公司
设计师：任萃
项目地点：台湾台北
项目面积：920 平方米
摄影：卢震宇
主要材料：洗石子崁铜条、线板漆平光黑漆、明镜、磁砖、纯白人造石

ZhongLi Teahouse Club

中莉茗茶会所

Design agency: Fujian Donny Decorative Engineering & Design Co., Ltd.
Designer: Chuan-dao Li
Location: Fuzhou, Fujian
Area: 360m²
Photography: Ling-yu Li

设计单位：福建东道建筑装饰设计有限公司
设计师：李川道
项目地点：福建福州
项目面积：360 平方米
摄影：李玲玉

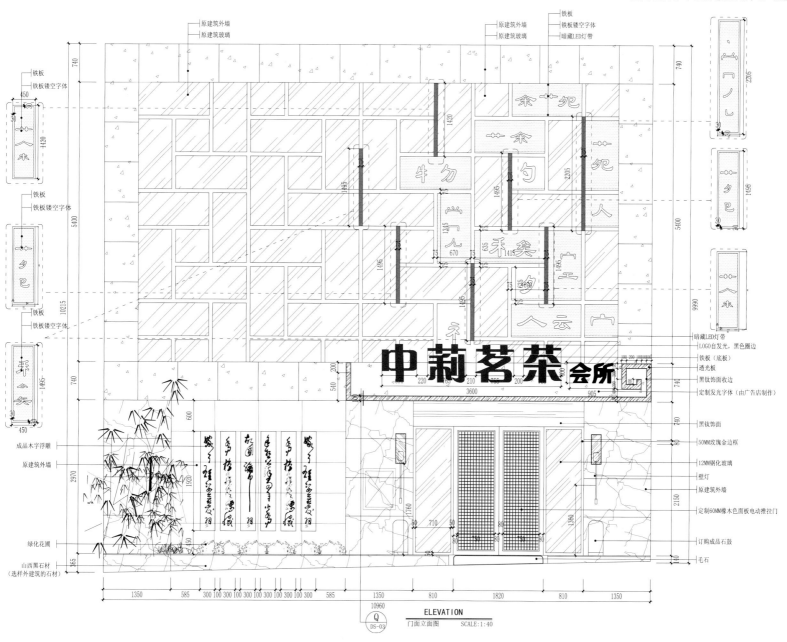

Xi Ming Reception Club

熹茗接待会所

Design agency: Daohe Design	设计单位：道和设计
Designer: Xiong Gao	设计师：高雄
Location: Fujian	项目地点：福建
Photography: Ling-yu Li	摄影：李玲玉
Main materials: white stone coating, white painted glass, coffee ashtree veneer, Mongolia black foiled stone	主要材料：白色真石漆、白色烤漆玻璃、水曲柳木饰面染咖色、蒙古黑火烧石

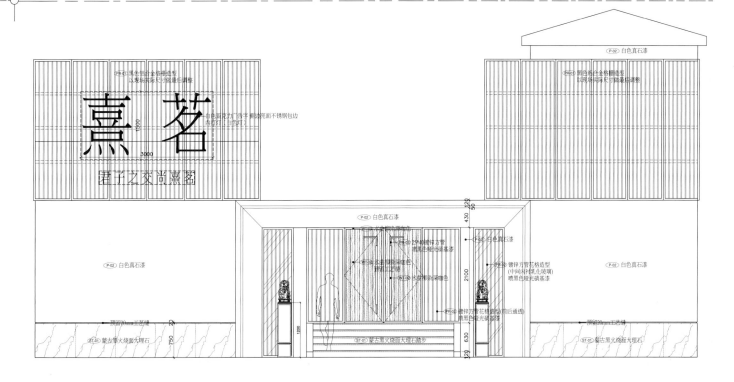

Hi Fit Gym

海菲特健身房

Design agency: XYI Design Consulting Co., Ltd.
Designer: Mac Huang
Location: Qingdao, Shandong
Area: 780m²
Photography: Tao Wang

设计单位：隐巷设计顾问有限公司
设计师：黄士华
项目地点：山东青岛
项目面积：780 平方米
摄影：王涛

Faces Plus Skin Spa

Faces Plus 美容会所

Design agency: AX Design Group
Location: San Francisco, California, USA
Area: 610m²

设计单位：大器联合室内装修设计有限公司
项目地点：美国加州旧金山
项目面积：610 平方米

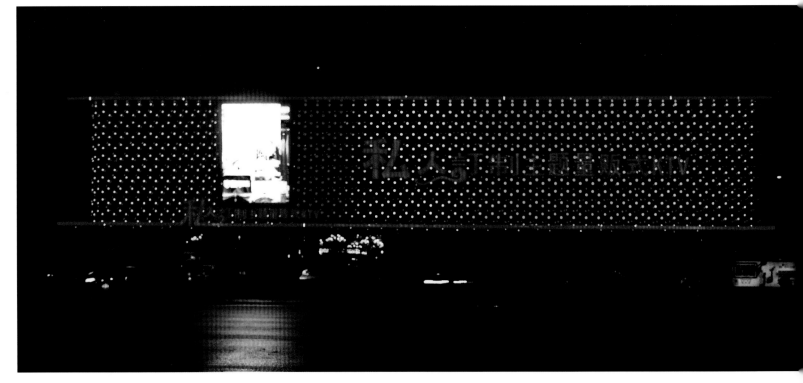

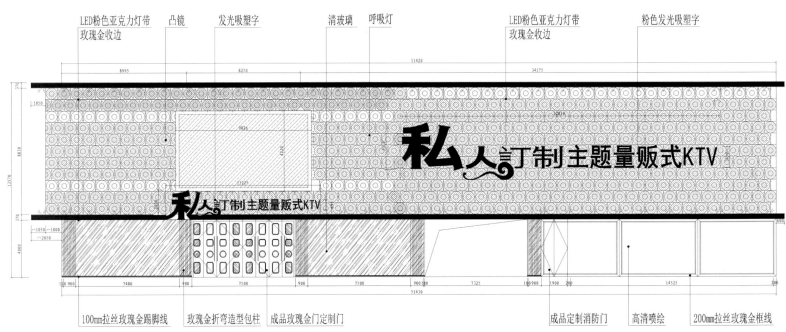

Private Costumed KTV

私人订制主题量贩 KTV

Design agency: SPACE³ Design Studio
Designer: Lin Xu, Di-long Ma, Hong-bo Wang
Location: Shenyang, Liaoning
Area: 3000m²

设计单位：加拿大立方体设计事务所
设计师：徐麟、马狄龙、王洪博
项目地点：辽宁沈阳
项目面积：3000 平方米

Yakitasty BBQ & Bar
串亭烧烤居酒屋

Design agency: Yuejie Design Company 设计单位：悦界设计
Designer: Ming Guo, Ge Ye 主创设计：郭明、叶格

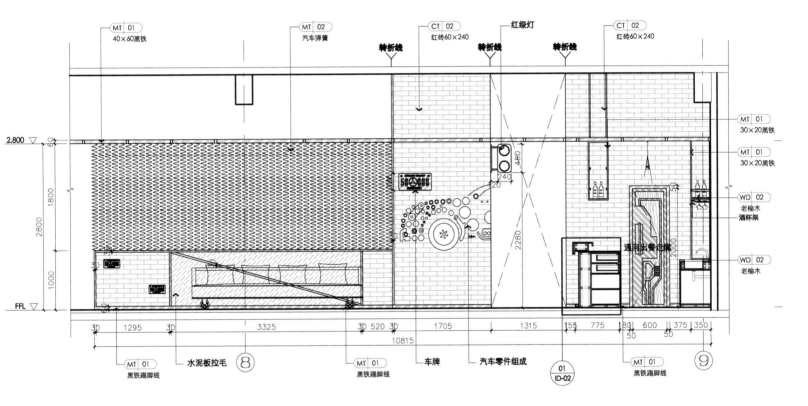

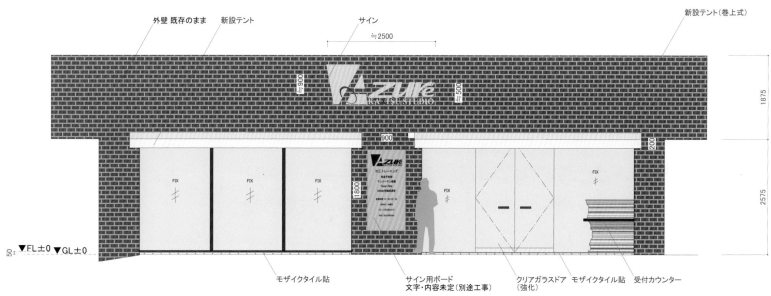

Azure Kaatsu Studio

Azure 健身会所

Design agency: Matsuya Art Works. Co., Ltd.
Designer: Tetsuya Matsumoto
Location: Himeji, Japan
Photography: Toshiyuki Nishimatsu

设计单位：松屋艺术设计有限公司
设计师：Tetsuya Matsumoto
项目地点：日本姬路市
摄影：Toshiyuki Nishimatsu

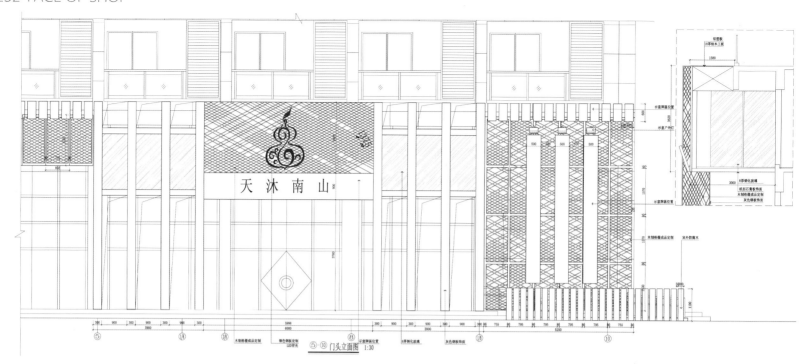

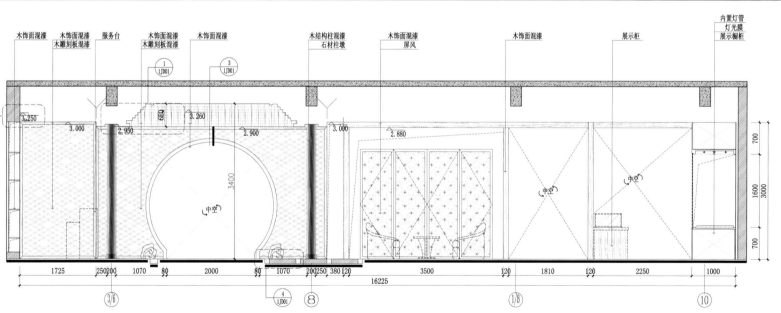

天沐南山中医养生会所

Timenasa Health-maintenance Club

Design agency: China Academy of Art-Shegu Design
Designer: Yin-qiu Xie
Location: Jiangsu

设计单位：中国美术学院设谷设计事务所
设计师：谢银秋
项目地点：江苏

Moon Beside the Lake Reception

唐乾明月接待会所

Design agency: Daohe Design
Designer: Xiong Gao
Location: Fujian
Photography: Kai Shi, Ling-yu Li
Main materials: tile, ecological wood

设计单位：道和设计
设计师：高雄
项目地点：福建
摄影：施凯、李玲玉
主要材料：瓷砖、生态木

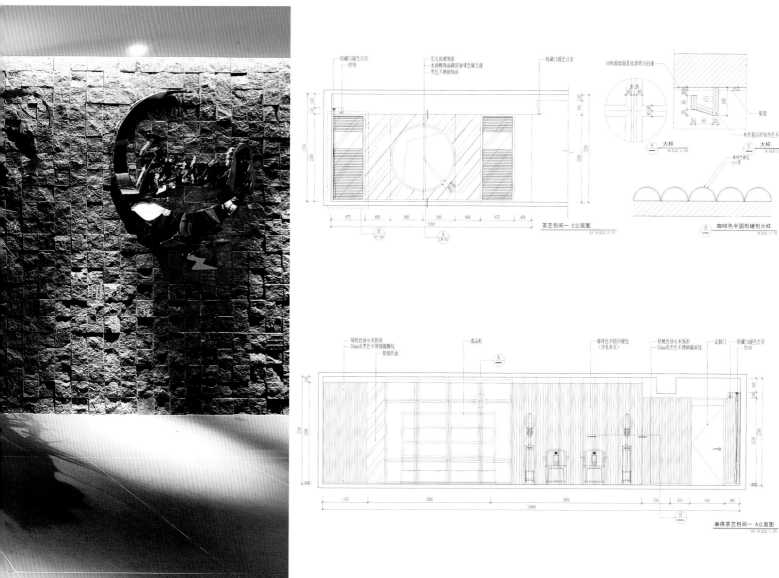

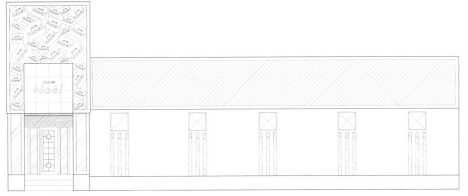

NOEL Membership Club

诺爱日式会员制俱乐部

Designer: Jian Zhang
Location: Dalian, Liaoning
Area: 1200m²
Main materials: marble, carpet, LED, fiber-optical, mirror, soft packing

设计师：张健
项目地点：辽宁大连
项目面积：1200平方米
主要材料：大理石、地毯、LED、光纤、镜面、软包

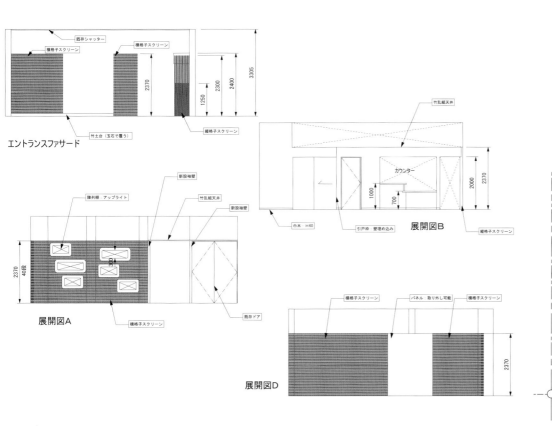

Refre 按摩馆

Design agency: Matsuya Art Works. Co., Ltd.
Designer: Tetsuya Matsumoto
Location: Japan
Photography: Toshiyuki Nishimatsu

设计单位：松屋艺术设计有限公司
设计师：Tetsuya Matsumoto
项目地点：日本
摄影：Toshiyuki Nishimatsu

Shanghai Palace of Moxa Flagship Shop

东方博艾馆上海旗舰店

Design agency: Shanghai ShanXiang Architectural Design Co.,Ltd.
Designer: Shan-xiang Wang
Location: Shanghai
Area: 505m²
Photography: Wen-jie Hu
Main material: oak, Mongolian scotch pine, parquet, volcanic rock, grey tile, pebble

设计单位：上海善祥建筑设计有限公司
设计师：王善祥
项目地点：上海
项目面积：505 平方米
摄影：胡文杰
主要材料：橡木、樟子松、实木复合地板、火山岩、青瓦、卵石

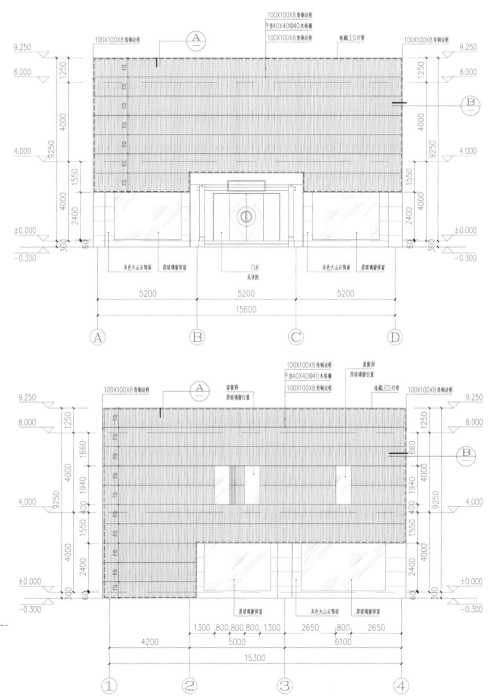

花想容：美容美发
Pretty Beauty: Beauty Salon

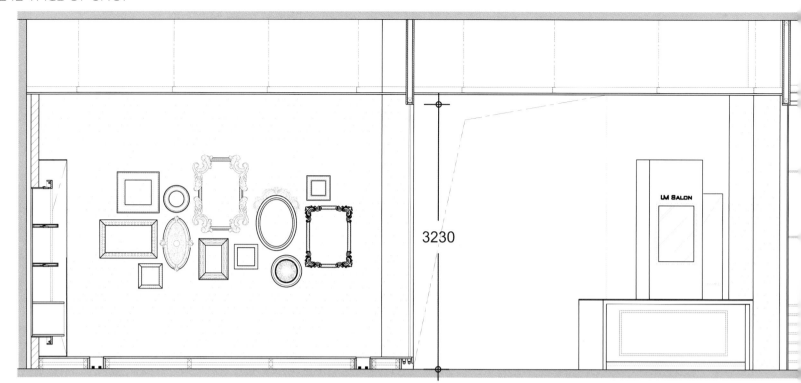

UM Salon
UM 发型概念店

Design agency: AS Design Service Limited
Designer: Four Lau, Sam Sum, Twiggy Yau
Location: Shenzhen, Guangdong
Area: 228m²
Photography: Sing Studio by Sum Sing
Materials: Stainless Steel Laser-cut Pattern, Back Paint Glass in Black Color, Clear Tempered Glass in Black Color, Matt PU Painting

设计单位：AS Design Service Limited
设计师：刘盛科、沈浩梁、游嘉欣
项目地点：广东深圳
项目面积：228 平方米
摄影：Sing Studio by Sum Sing
主要材料：不锈钢激光切割铁花、黑底油玻璃、强化黑玻璃、哑面聚氨酯涂料

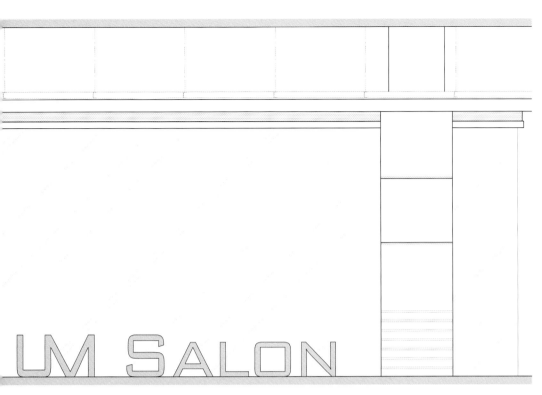
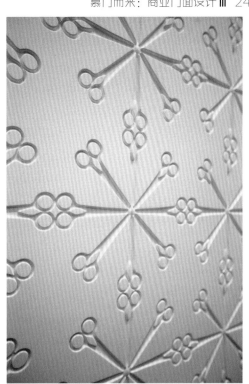

Neilsmap
Neilsmap 美甲店

Design agency: SYK Interior Design Co.,Ltd.　　设计单位：拾雅客空间设计
Designer: Janus　　设计师：许炜杰
Location: Taipei, Taiwan　　项目地点：台湾台北

Zi Heng
姿恒

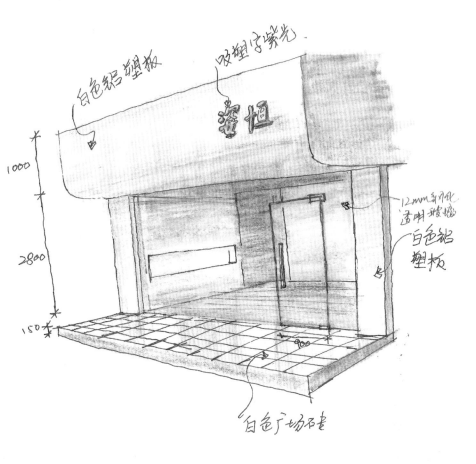

Designer: Richard Hong
Location: Shanwei, Guangdong
Photography: Wei-jin Ou

设计师：康世畅
项目地点：广东汕尾
摄影：欧伟金

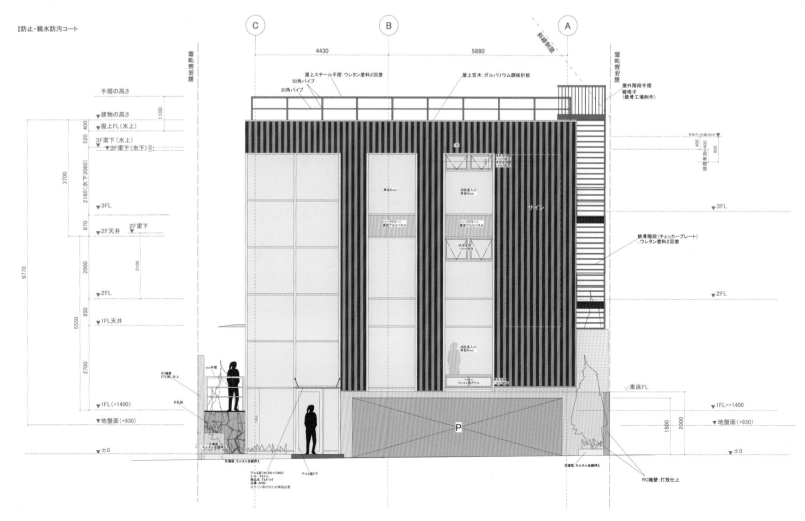

Musée Hair & Make

Musée 美发馆

Design agency: Matsuya Art Works. Co., Ltd.	设计单位：松屋艺术设计有限公司
Designer: Tetsuya Matsumoto	设计师：Tetsuya Matsumoto
Location: Kobe, Japan	项目地点：日本神户市
Photography: Toshiyuki Nishimatsu	摄影：Toshiyuki Nishimatsu

In Health & Beauty Nail Salon

In 美甲店

Design agency: Wang Feng-bo Design Studio
Designer: Dong Meng
Location: Beijing

设计单位：北京王凤波设计机构
设计师：孟冬
项目地点：北京

Back Ground Spa Club

逅台美容Spa会所

Design agency: HK ShengJiaYou Interior Design Co., Ltd.	设计单位：香港绳家友国际设计机构
Designer: Dolphin Sheng	设计师：绳家友
Location: Fenghua, Ningbo, Zhejiang	项目地点：浙江宁波奉化
Area: 1580m²	项目面积：1580平方米
Photography: Lina	摄影：李影

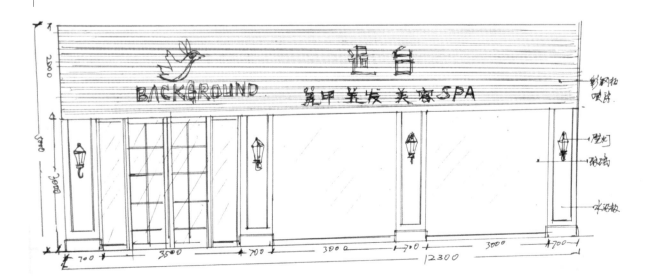

Lady Care Cosmetic
女士美妆店

Design agency: Tiko Interiors Ltd.
Designer: Matthew Tong
Location: Hong Kong
Area: 45m²

设计单位：天皓设计工程有限公司
设计师：唐万强
项目地点：香港
项目面积：45 平方米

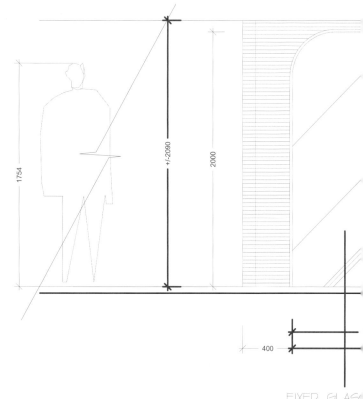

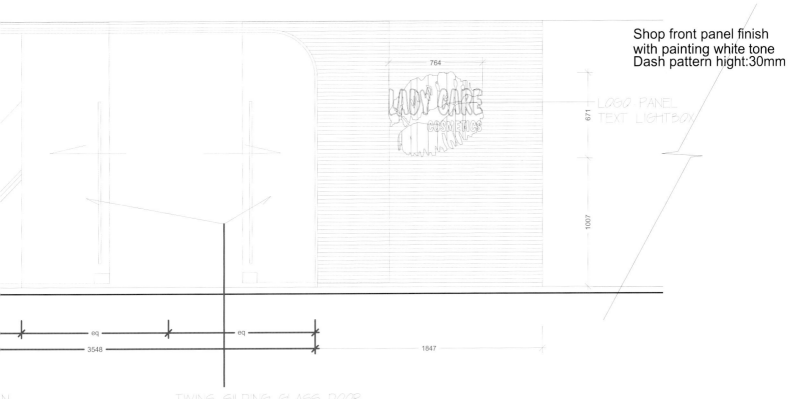

Shop front panel finish with painting white tone
Dash pattern hight:30mm

珠宝配饰：黄金玉器
Jewelry Accessories:Gold,Jade

Premiera, Bangkok, Thailand

曼谷 Premiera 珠宝专卖店

Design agency: Design Worldwide Partnership
Location: Bangkok, Thailand
Area: 200m²

设计单位：DWP 设计公司
项目地点：泰国曼谷
项目面积：200 平方米

Shimansky Jewelers

诗曼斯科珠宝专卖店

Design agency: HEAD Architecture and Design
Location: Macau

设计单位：HEAD建筑设计有限公司
项目地点：澳门

Joylong Jewelry Store

金银楼珠宝店

Designer: Zhi-fei Zhang
Location: Taizhou, Zhejiang
Area: 110m²
Photography: Zhi-fei Zhang
Main materials: cherry, stainless steel, tawny mirror, soft roll, cord, wooden flooring, tile, paint

设计师：张智飞
项目地点：浙江台州
项目面积：110 平方米
摄影：张智飞
主要材料：美国红樱桃、电镀不锈钢、灰茶镜、软膜、线帘、木地板、地砖、涂料

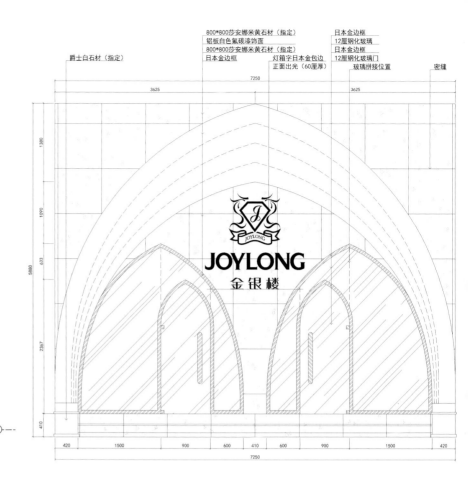

Mont Blanc
万宝龙专卖店

Design agency: Design Worldwide Partnership
Location: Bangkok, Thailand

设计单位：DWP 设计公司
项目地点：泰国曼谷

Heyuan China Gold Store
中国黄金河源店

Designer: Zhi-fei Zhang
Location: Heyuan, Guangdong
Area: 550m²
Main materials: stainless steel, pre-straining glass, wired glass, metal net, leather, prismatic crystal mirror, serpeggiante, wallcloth, teakwood

设计师：张智飞
项目地点：广东河源
项目面积：550 平方米
主要材料：电镀不锈钢、钢化玻璃、夹丝玻璃、金属网、皮料、棱柱水晶镜、木纹石、墙布、柚木

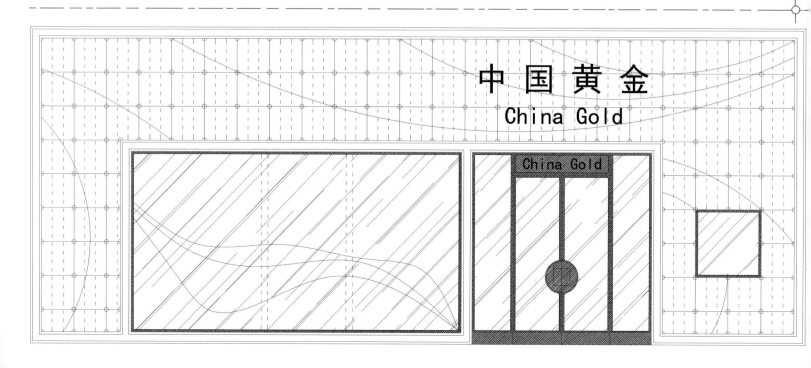

Broad Optical
博大眼镜

Designer: Richard Hong
Location: Shenzhen, Guangdong
Photography: Wei-jin Ou

设计师：康世畅
项目地点：广东深圳
摄影：欧伟金

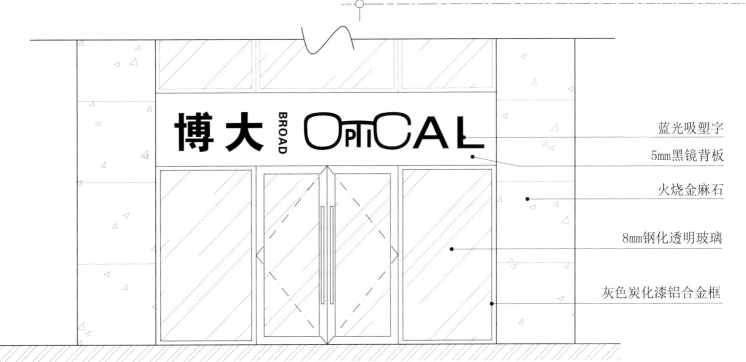

- 蓝光吸塑字
- 5mm黑镜背板
- 火烧金麻石
- 8mm钢化透明玻璃
- 灰色炭化漆铝合金框

Optique Ampere Store
Optique Ampere 眼镜店

Designer: Cyrille Druart　　设计师：西里·杜拉特
Location: France　　项目地点：法国

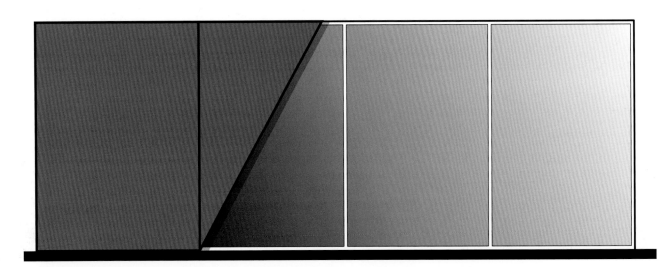

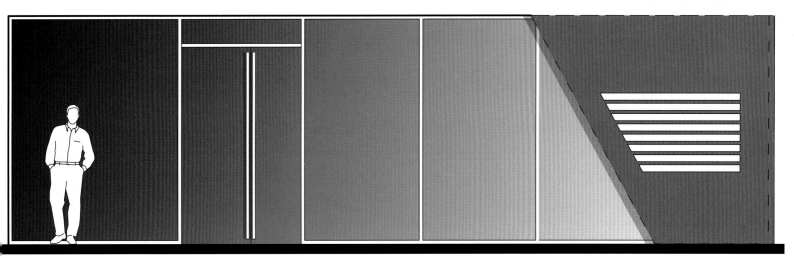

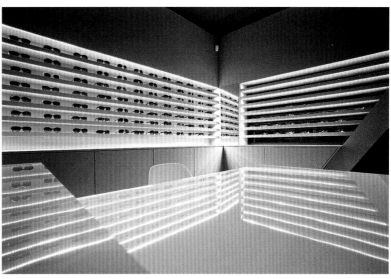
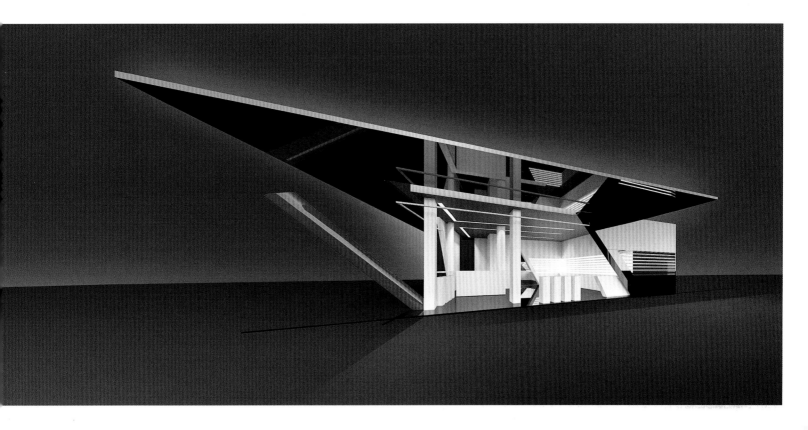

Nixon Store in London
伦敦尼克松手表专卖店

Design agency: Checkland Kindleysides
Location: London, UK
Area: 24m²

设计单位：Checkland Kindleysides 设计顾问公司
项目地点：英国伦敦
项目面积：24 平方米

Nixon Store in Paris
巴黎尼克松手表专卖店

Design agency: Checkland Kindleysides
Location: Paris, France
Area: 77m²

设计单位：Checkland Kindleysides 设计顾问公司
项目地点：法国巴黎
项目面积：77 平方米

Ops Object Store

Ops Object 专卖店

Design agency: Fabio Caselli Design
Location: Basilea, Switzerland

设计单位：法比奥·卡塞设计工作室
项目地点：瑞士巴塞尔

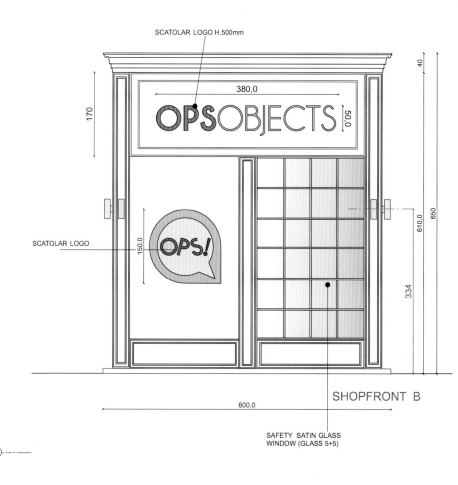

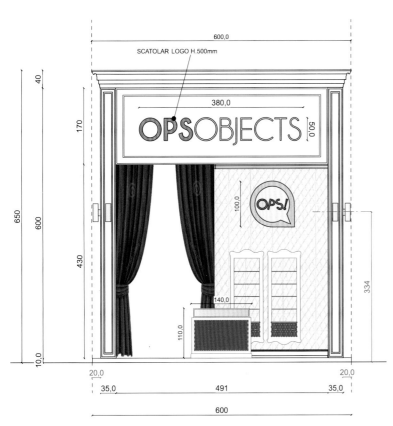
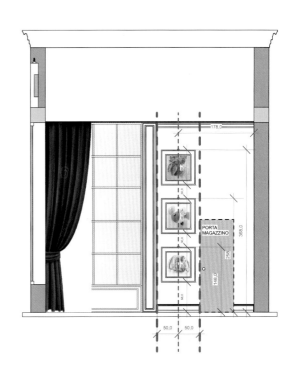

爱家：艺术家居
Love Home: Artistic Furnishing

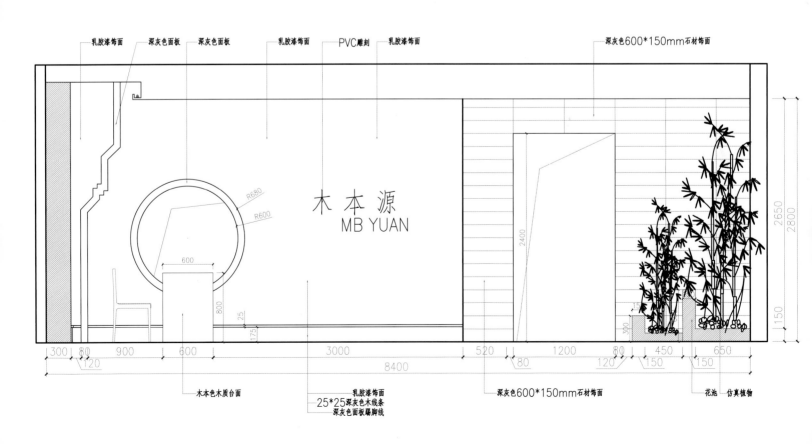

Mu Ben Yuan Chinese Style Bathroom Products Flag-store

木本源中式卫浴旗舰展厅

Design agency: Chengdu S.X. Design
Location: Chengdu, Sichuan
Area: 160m²

设计单位：成都私享室内设计有限公司
项目地点：四川成都
项目面积：160 平方米

Kitchen Aid Flagship Store

佳霆厨具旗舰店

Design agency: XYI Design Consultants Limited
Designer: Eve Yuan
Location: Zhong-shan district, Taipei
Area: 72m²
Photography: Ji-shou Wang

设计单位：隐巷设计顾问有限公司
设计师：袁筱媛
项目地点：台北中山区
项目面积：72平方米
摄影：王基守

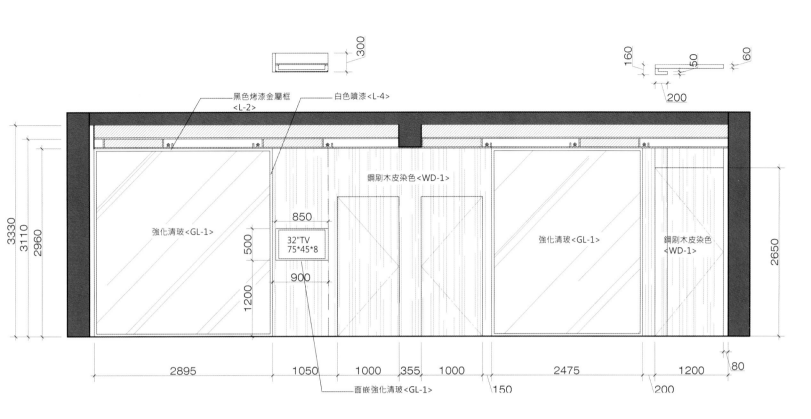

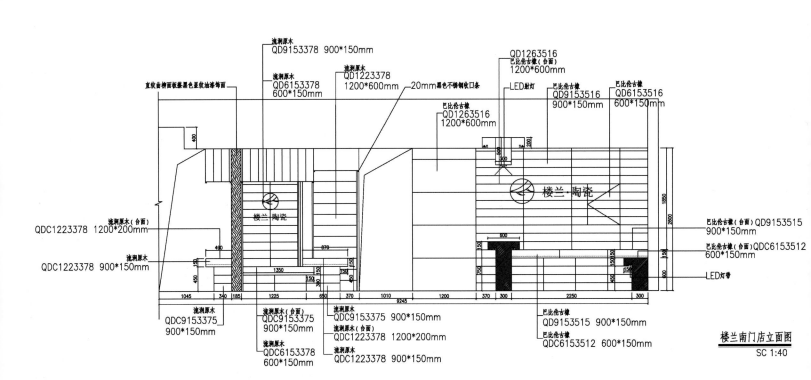

Lola Ceramics
楼兰陶瓷

Design agency: Chengdu S.X. Design
Location: Chengdu, Sichuan
Area: 110m²

设计单位：成都私享室内设计有限公司
项目地点：四川成都
项目面积：110平方米

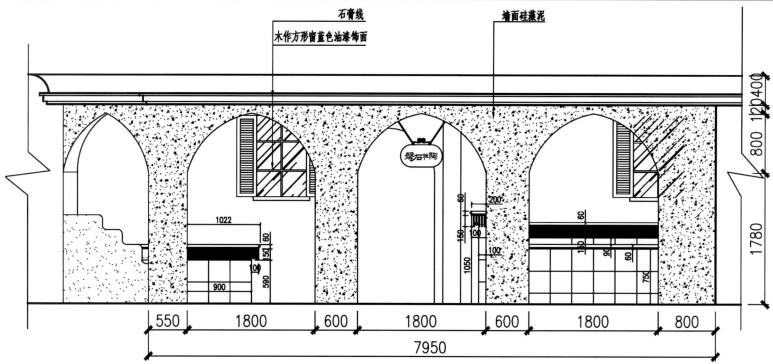

Pan Shi Ya Tao

磐石雅陶

Design agency: Chengdu S.X. Design
Location: Chengdu, Sichuan
Area: 160m²

设计单位：成都私享室内设计有限公司
项目地点：四川成都
项目面积：160 平方米

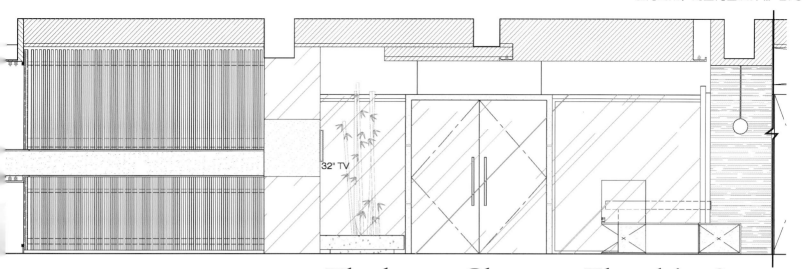

Zhuhuan Cleanup Flagship Store

竹桓 Cleanup 旗舰店

Design agency: XYI Design Consultants Limited
Designer: Mac Huang, Carrie Meng, Eva Yuan
Location: Zhong-shan district, Taipei
Area: 333m²
Main materials: parasol-tree wood panel, black iron panel, HPL, Indian Black marble, Ariston marble, ultra white forsted glass, black mirror

设计单位：隐巷设计顾问有限公司
设计师：黄士华、孟羿彣、袁筱媛
项目地点：台北中山区
项目面积：333 平方米
主要材料：钢刷梧桐实木板、黑铁板、美耐板、印度黑理石、雪白银狐理石、超白喷砂玻璃、黑镜

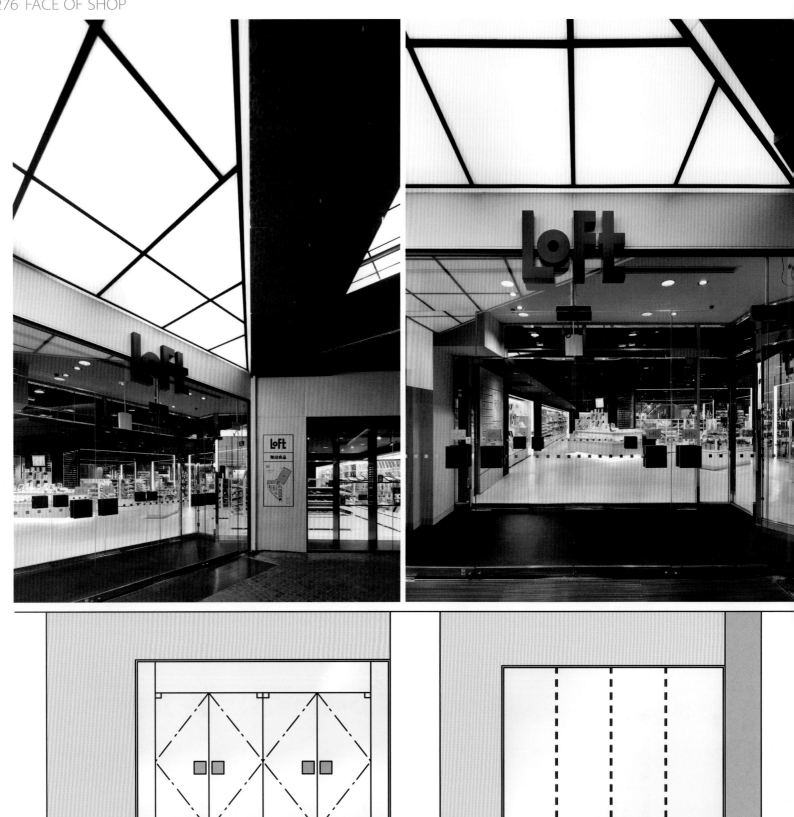

LOFT Shibuya
涉谷LOFT百货

Design agency: Moment
Designer: Hisaaki Hirawata, Tomohiro Watabe
Location: Shibuya, Tokyo, Japan
Area: 930m²
Photography: Nacasa & Partners Inc.

设计单位：Moment 设计事务所
设计师：Hisaaki Hirawata, Tomohiro Watabe
项目地点：日本东京涉谷区
项目面积：930 平方米
摄影：Nacasa & Partners 摄影工作室

Akomeya Tokyo

Akomeya Tokyo 米行

Design agency: Moment
Designer: Hisaaki Hirawata, Tomohiro Watabe
Location: Ginza Tokyo, Japan
Area: 600m²
Photography: Fumio Araki
Main materials: ceramic tile, stucco painting, wood wall

设计单位：Moment 设计事务所
设计师：Hisaaki Hirawata, Tomohiro Watabe
项目地点：日本东京银座
项目面积：600 平方米
摄影：Fumio Araki
主要材料：瓷砖、STUCCO 涂料、木墙

Forest in Xinxiang

新乡森林

Design agency: SAKO Architects
Designer: Sako Keiichiro, Fujii Yoko, Takaoka Yuji, Zhi-yong Liu
Location: Xinxiang, Henan
Area: 7000m²
Photography: Ruijing Photography
Main materials: glass facade, aluminium alloy window, LED screen, tile, gypsum board, stainless steel tube, wood, mirror

设计单位：SAKO 建筑设计工社
设计师：迫庆一郎、藤井洋子、高冈勇治、刘智勇
项目地点：河南新乡
项目面积：7000 平方米
摄影：锐景摄影
主要材料：玻璃幕墙、铝合金窗、LED 屏幕、瓷砖、石膏板、不锈钢管、木材、镜子

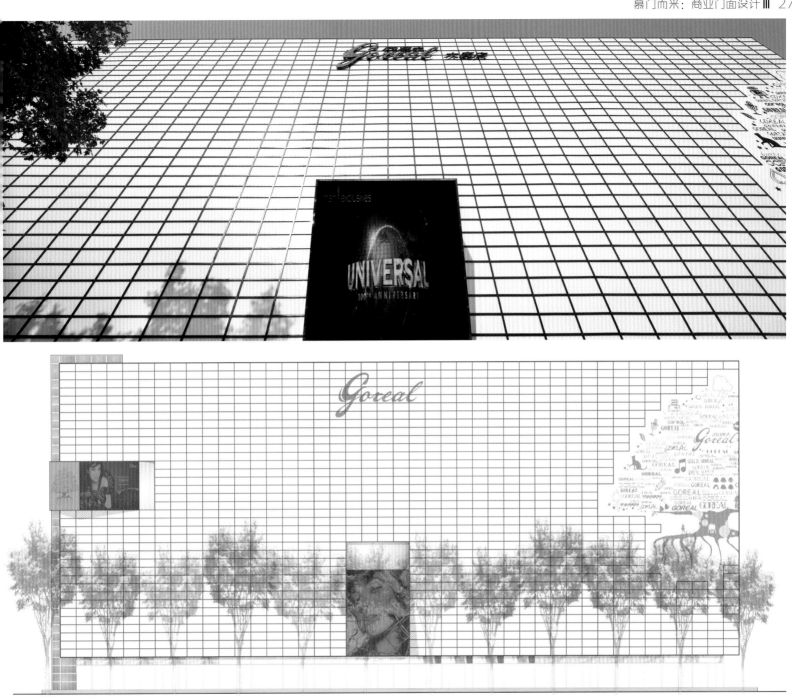
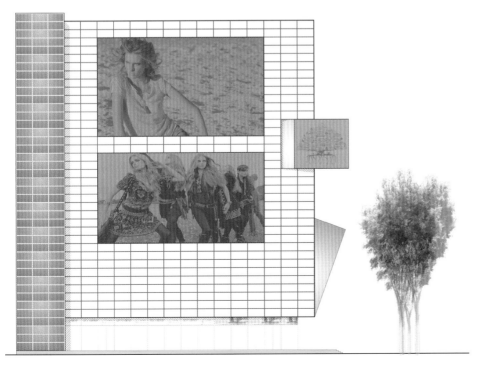

Kayanoya Tokyo

东京茅乃舍酱油店

Design agency: Kengo Kuma and Associates
Designer: Kengo Kuma
Location: Japan

设计单位：隈研吾建筑都市设计事务所
设计师：隈研吾
项目地点：日本

Van Gogh Ceramics 1583

梵高陶瓷 1583

Design agency: Chengdu S.X. Design
Designer: Jun-feng Hu
Location: Chongqing
Area: 150m²

设计单位：成都私享室内设计有限公司
设计师：胡俊峰
项目地点：重庆
项目面积：150 平方米

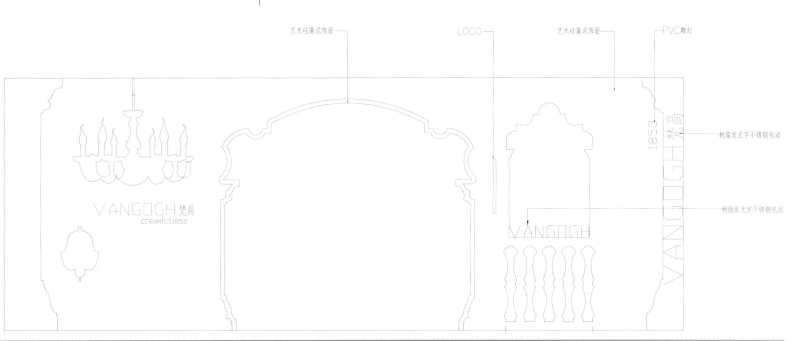

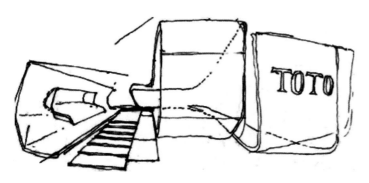

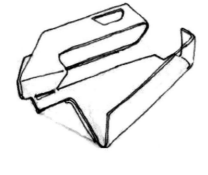
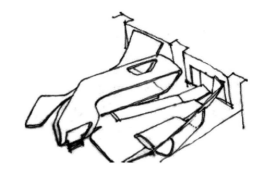

TOTO Xinyi Flagship Store
东陶卫浴信义旗舰店

Design agency: AX Design Group
Location: Taipei, Taiwan
Area: 53m²

设计单位：大器联合室内装修设计有限公司
项目地点：台湾台北
项目面积：53 平方米

Rachily Bella
瑞奇贝拉家具馆

Design agency: Ganna Design Studio
Designer: Shih-jie Lin, Ting-liang Chen
Location: Taipei, Taiwan
Area: 105m²

设计单位：甘纳空间设计
设计师：林仕杰、陈婷亮
项目地点：台湾台北
项目面积：105 平方米

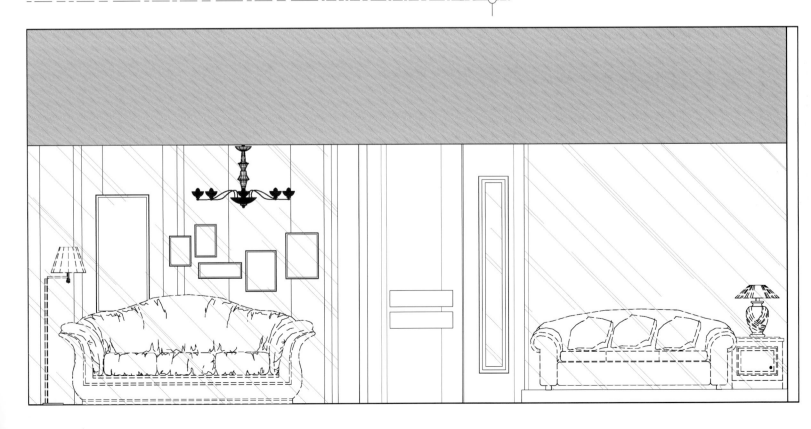

KEEY Lighting Exhibition Hall
企一照明展厅

Design agency: JiaYi Interior Design Co., Ltd.
Designer: Hong-bin Wang
Location: Zhongshan, Guangdong
Area: 70m²
Main materials: imitated marble tile, rose gold fluorocarbon paint, oil painting, latex paint

设计单位：佳易室内装饰设计有限公司
设计师：王鸿斌
项目地点：广东中山
项目面积：70 平方米
主要材料：仿大理石砖、高档玫瑰金氟碳漆、油画框线条、乳胶漆

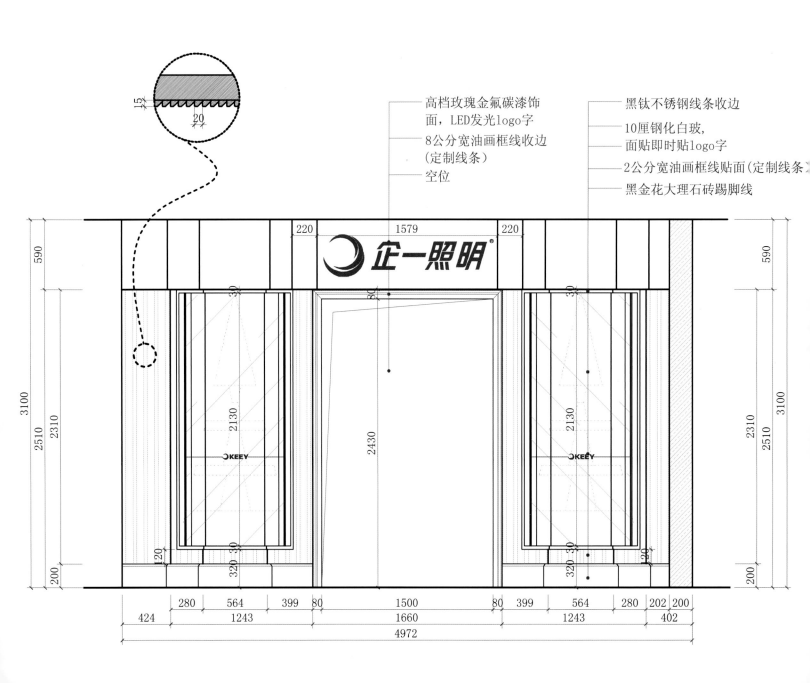

KEEY Lighting Store

企一照明专卖店

Design agency: JiaYi Interior Design Co., Ltd.
Designer: Hong-bin Wang, Jin-sheng Luo
Location: Zhongshan, Guangdong
Area: 390m²
Main materials: wallpaper, artificial stone, champagne-white marble, white brushing lacquer

设计单位：佳易室内装饰设计有限公司
设计师：王鸿斌、罗井生
项目地点：广东中山
项目面积：390 平方米
主要材料：墙纸、人造石、香槟白大理石、白色手扫漆

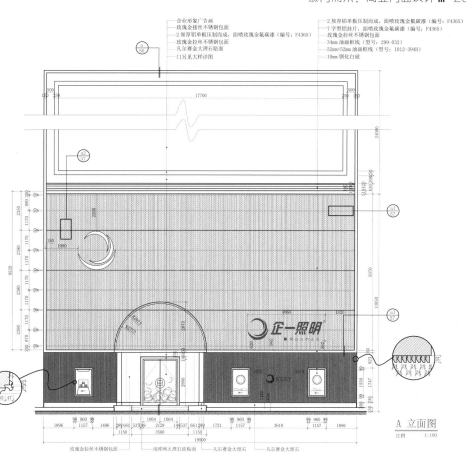

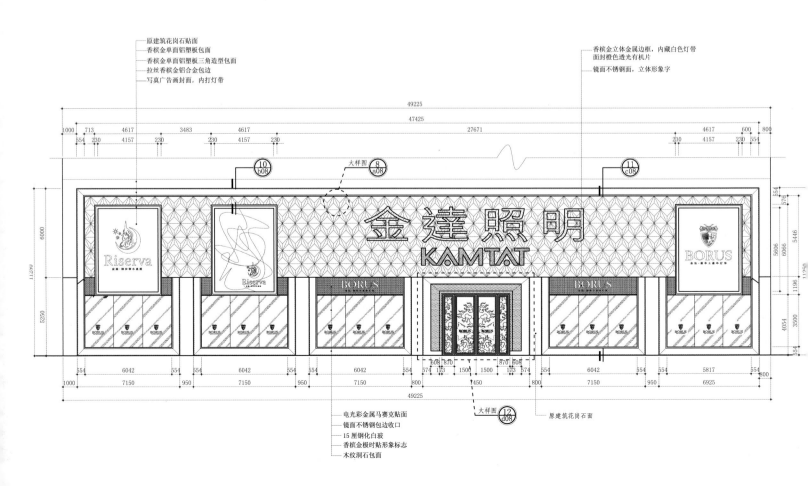

Kamtat
Lighting Life Mall

金达照明品牌灯饰生活馆

Design agency: JiaYi Interior Design Co., Ltd.
Designer: Hong-bin Wang, Wei-guo Jia, Jin-sheng Luo
Location: Dongguan, Guangdong
Area: 5000m²

设计单位：佳易室内装饰设计有限公司
设计师：王鸿斌、贾伟国、罗井生
项目地点：广东东莞
项目面积：5000 平方米

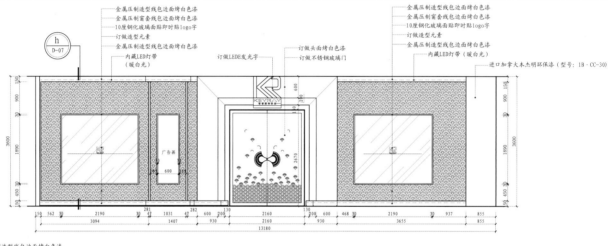

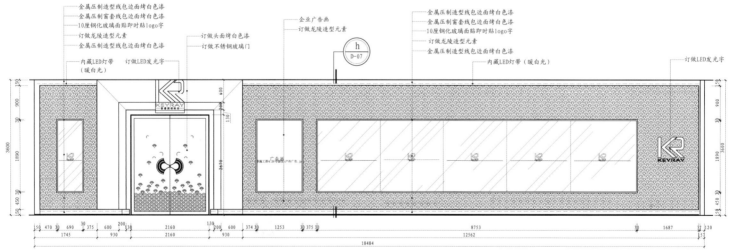

Keyray Lighting Brand Store

智睿照明科技品牌形象店

Design agency: JiaYi Interior Design Co., Ltd.	设计单位：佳易室内装饰设计有限公司
Designer: Hong-bin Wang	设计师：王鸿斌
Location: Zhongshan, Guangdong	项目地点：广东中山
Area: 240m²	项目面积：240平方米
Main materials: laminated wooden flooring, painting, marble, baking metal panel, veneer	主要材料：复合木地板、进口涂料、大理石、金属板烤漆、饰面板

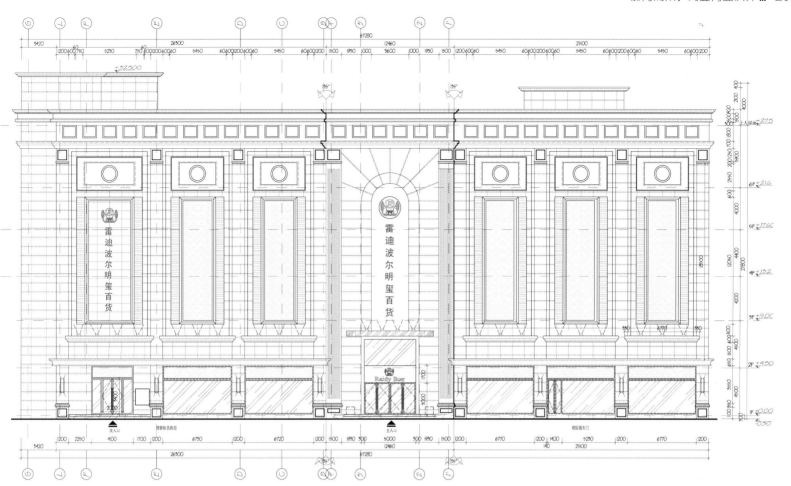

Raidy Boer MingXi Shopping Mall

雷迪波尔明玺百货

Design agency: ShangDa Design Co., Ltd.
Designer: Hai-tao Liu
Location: Chengdu, Sichuan
Area: 20000m²

设计单位：SDD 上达国际
设计师：刘海涛
项目地点：四川成都
项目面积：20000 平方米

De Rucci Flagship Store
慕思旗舰店

Design agency: Rocky Design HK Assiociates
Designer: Rocky Chan
Location: Dongguan, Guangdong
Area: 3000m²
Main materials: stone, wood veneer, glass, wallpaper, cloth, flooring, stainless steel

设计单位：陈飞杰香港设计事务所
设计师：陈飞杰
项目地点：广东东莞
项目面积：3000平方米
主要材料：石材、木饰面、玻璃、壁纸、布艺、地板、不锈钢

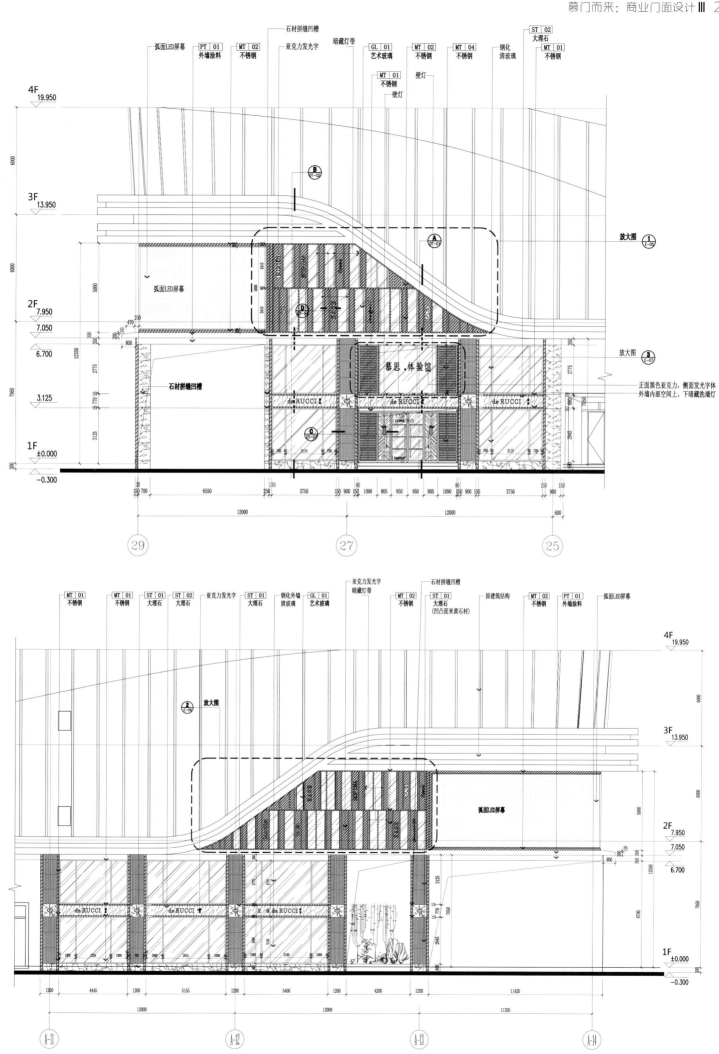

De Rucci Experience Shop

慕思体验馆

Design agency: Rocky Design HK Assiociates
Designer: Rocky Chan
Location: Dongguan, Guangdong
Main materials: stone, wood veneer, glass, wallpaper, cloth, flooring, stainless steel

设计单位：陈飞杰香港设计事务所
设计师：陈飞杰
项目地点：广东东莞
主要材料：石材、木饰面、玻璃、壁纸、布艺、地板、不锈钢

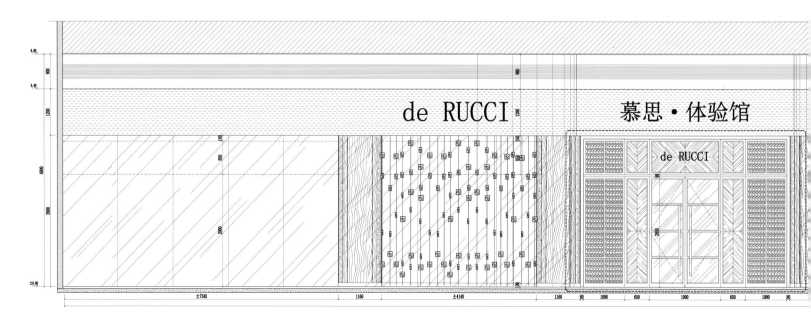

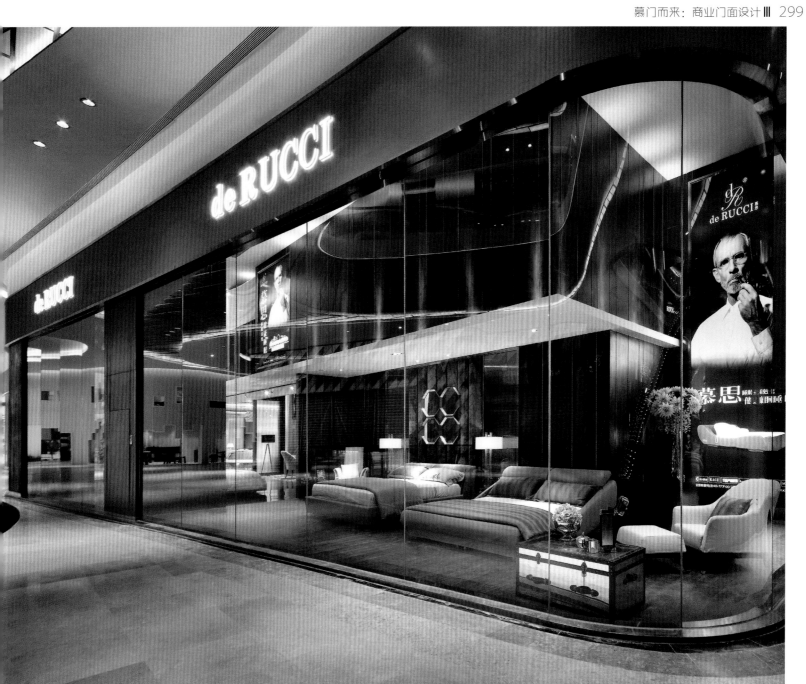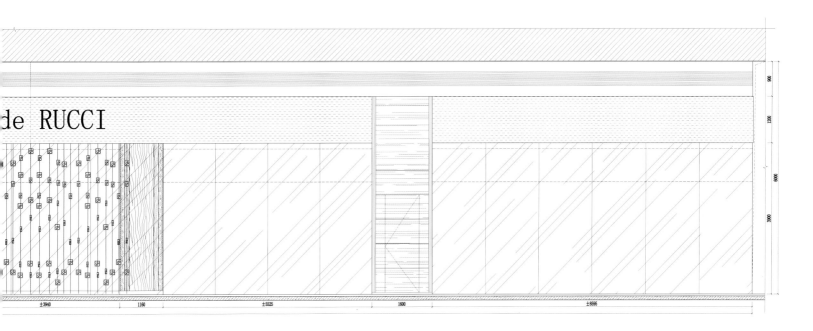

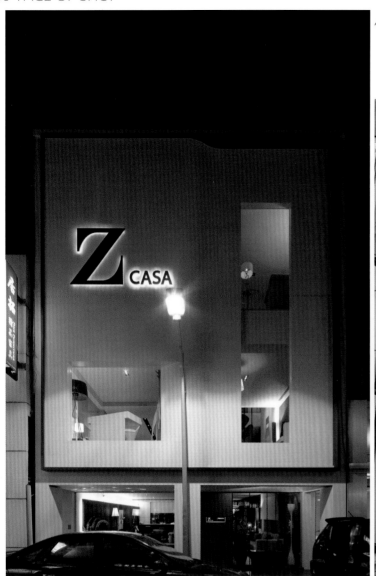

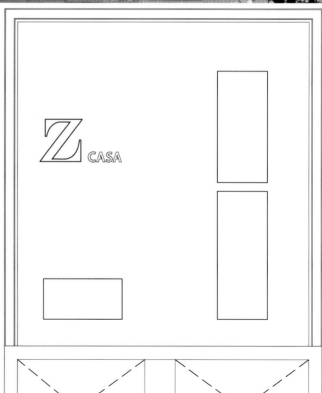

Wenchang Z Casa Store

Z Casa 台北文昌店

Design agency: AX Design Group
Location: Taipei, Taiwan
Area: 390m²

设计单位：大器联合室内装修设计有限公司
项目地点：台湾台北
项目面积：390 平方米

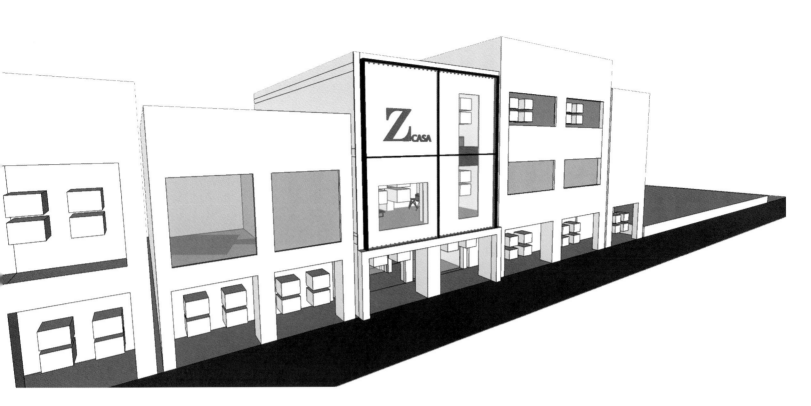

LOHAS Smart Home
乐活智能家居会所

Designer: Jian Zhang
Location: Dalian, Liaoning
Area: 350m²
Main materials: granite, hollow brick, marble translucent piece, white crocodile leather, tencel plate

设计师：张健
项目地点：辽宁大连
项目面积：350 平方米
主要材料：花岗岩、空心砖、云石透光片、白色鳄鱼皮、天丝板

Oriental Yard

东方大院

Design agency: Daohe Design　　设计单位：道和设计
Designer: Xiong Gao　　设计师：高雄
Location: Fuzhou, Fujian　　项目地点：福建福州

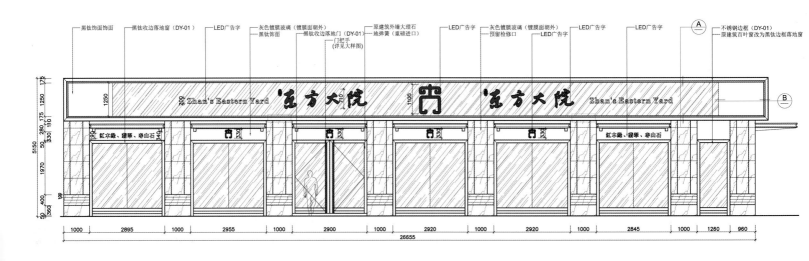

外观A立面施工图

- 黑钛饰面 18厘板基层
- 暗藏T5白色灯带（两条）
- 白色烤漆玻璃面乳化 18厘板基层
- LED广告字
- LED广告字电源垂直引入
- 灰色镀膜玻璃（镀膜面朝外）

A剖面图
A2 SCALE:1:20

- 18厘板基层
- 白色烤漆玻璃面乳化
- 暗藏T5白色灯带（两条）
- 灰色镀膜玻璃（镀膜面朝外）
- 18厘板基层

- 黑钛饰面玻璃卡条（自攻螺丝固定）
- 黑钛饰面 18厘板基层
- 黑钛饰面 18厘板基层

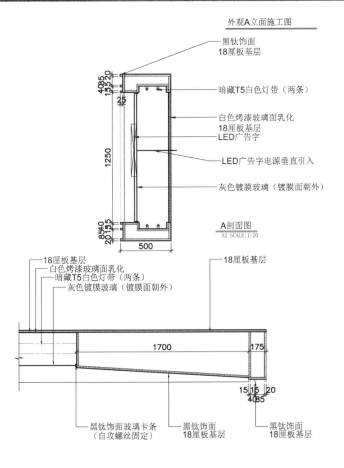

Aleybo Flagship Store

Aleybo 厨具用品旗舰店

Design agency: 5 Star Plus Retail Design
Designer: Bruce Li, Barbara Seidelmann, Ray Zhang
Location: Beijing
Area: 100m²
Photography: Bruce Li

设计单位：斐思达品牌设计咨询有限公司
设计师：Bruce Li, Barbara Seidelmann, Ray Zhang
项目地点：北京
项目面积：100平方米
摄影：Bruce Li

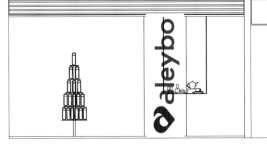

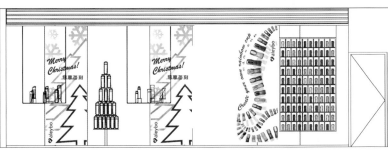

Ausnn the Australian Shop
澳牛牛澳洲生活馆

Design agency: Trenchant (Hangzhou) Design Co.,Ltd. 设计单位：杭州宣驰装饰设计有限公司
Designer: Qvent Xu 设计师：许立强
Location: Hangzhou, Zhejiang 项目地点：浙江杭州

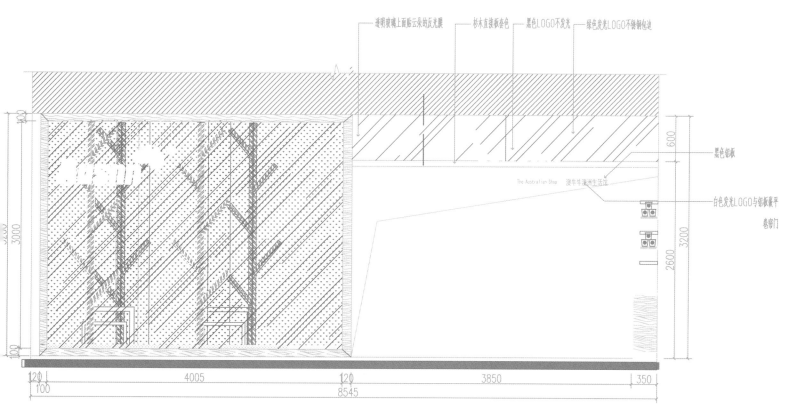

Merillat Flagship Shop

美睿厨房亚洲旗舰店

Design agency: Guangzhou VSA Brand Design Co., Ltd.
Designer: MAKE
Location: Guangzhou, Guangdong
Main materials: microlite, tungsten steel, rustic tile, wood, red brick, wood flooring

设计单位：广州意尚品牌策划有限公司
设计师：刘不丹
项目地点：广东广州
主要材料：微晶石、钨钢、复古砖、原木、红砖、实木地板

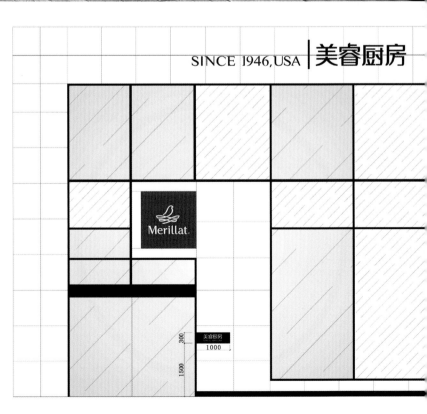

Su Mali Concept Shop Creative & Fashion

苏马利时尚创意概念店

Design agency: SYK Interior Design Co.,Ltd.	设计单位：拾雅客空间设计
Designer: Janus	设计师：许炜杰
Location: Taipei, Taiwan	项目地点：台湾台北
Area: 30m²	项目面积：30 平方米
Main materials: plastic flooring, cultural stone, painting, woodwork	主要材料：塑料地板、文化石、喷漆、木作

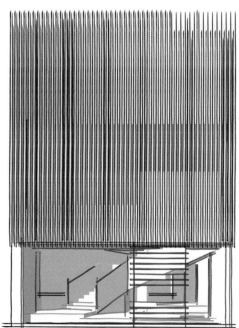

Golden • Yumu Designer's Lab
戈登隅木生活馆

Design agency: Hon Idea Design Studio
Designer: Beni Yeung
Location: Foshan, Guangdong
Area: 450m²
Photography: Beni Yeung

设计单位：硕瀚创意设计研究室
设计师：杨铭斌
项目地点：广东佛山
项目面积：450 平方米
摄影：杨铭斌

Jackes Doors Exhibition Hall

佳洁斯门业展厅

Designer: Zhi-min Liu
Location: Guangzhou, Guangdong
Area: 200m²

设计师：刘智铭
项目地点：广东广州
项目面积：200 平方米

聚焦：售楼处
Focus: Sales Office

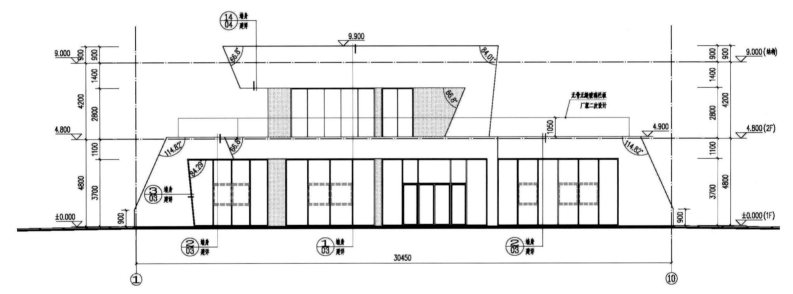

Huadi View Reception Center

华地远景接待中心

Design agency: XYI Design Consultants Limited
Designer: Mac Huang
Location: Rizhao, Shandong
Photography: Tao Wang
Main materials: Ariston, marble tile, grey oak wood, oak flooring, white finishing paint, ceramics tile

设计单位：隐巷设计顾问有限公司

设计师：黄士华

项目地点：山东日照

摄影：王涛

主要材料：雅士白理石、理石磁砖、橡木染灰木皮、橡木染灰地板、白色烤漆、陶瓷金属马赛克

Showcase Life Sales Center

金色领域售楼部

Design agency: Guangzhou C&C Design Co., Ltd.
Designer: Zheng Peng, Ze-kun Xie, Yong-xia Chen
Location: Foshan, Guangdong
Area: 800m²
Photography: IVY Photography&Production
Main materials: marble, baking finishing panel, latex painting, stainless steel, glass, fire-proof panel

设计单位：广州共生形态工程设计有限公司
设计师：彭征、谢泽坤、陈泳夏
项目地点：广东佛山
项目面积：800平方米
摄影：IVY. 蔓摄影
主要材料：大理石、烤漆板、乳胶漆、不锈钢、玻璃、防火板

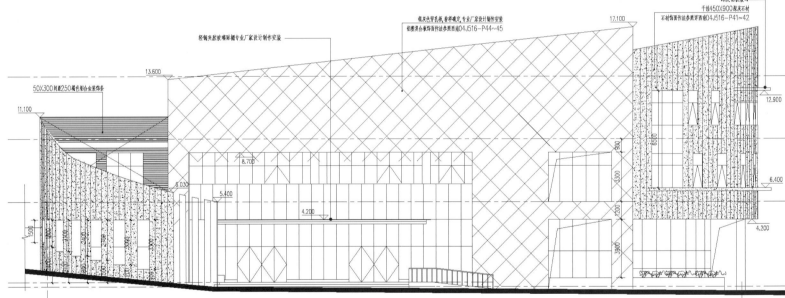

The Scene of Xishan Sales Center

溪山御景营销中心

Design agency: XYI Design Consultants Limited
Designer: Mac Huang
Location: Guiyang, Guizhou
Area: 600m²
Main materials: Ariston, Travertine, Sandstone, artificial stone, mental mosaic, ceramics mosaic, baking painted board, oak wood, wool carpet, mirror, white baking finishing glass

设计单位：隐巷设计顾问有限公司
设计师：黄士华
项目地点：贵州贵阳
项目面积：600 平方米
主要材料：雅士白、白洞石、砂岩大理石、人造石、金属马赛克、陶瓷马赛克、镜面烤漆板、橡木皮、羊毛地毯、明镜、白色烤漆玻璃

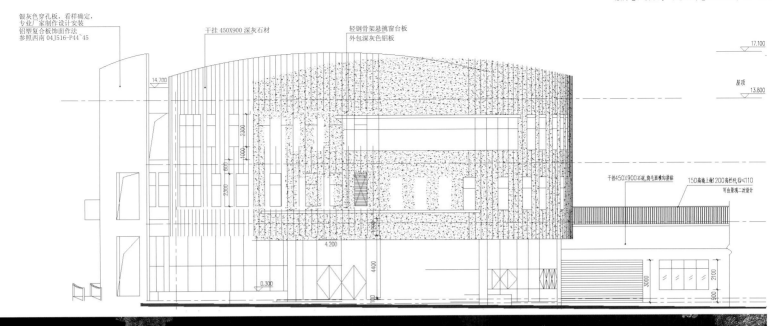

Vanke Park Avenue Reception Center

万科公园里接待中心

Design agency: XYI Design Consultants Limited
Designer: Mac Huang, Yi-jun Ding
Location: Ningbo, Zhejiang
Area: 1000m²
Main materials: White Marble, Venato Carrara, black mirror finishing stainless steel, Oak grille, white baking finishing panel, white artificial stone

设计单位：隐巷设计顾问有限公司
设计师：黄士华、丁义军
项目地点：浙江宁波
项目面积：1000 平方米
主要材料：白大理石、中花白大理石、黑色镜面不锈钢、染黑橡木格栅、白色烤漆板、白色人造石

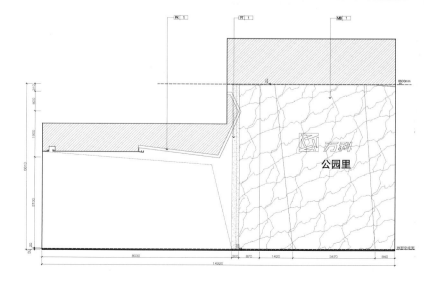

Riverside Manor
九夏云水售楼中心

Design agency: Simon Chong Design Consultants Ltd. **Designer:** Simon Chong, Amy Du **Location:** Kunming, Yunnan **Area:** 1800m²	设计单位：郑树芬设计事务所 设计师：郑树芬、杜恒 项目地点：云南昆明 项目面积：1800 平方米

Wuzhou Lakeside Sales Center
五洲家园售楼处

Design agency: Space Image Design & Decoration Co., Ltd.
Location: Zhuhai, Guangdong
Area: 800m²

设计单位：空间印象建筑装饰设计有限公司
项目地点：广东珠海
项目面积：800 平方米

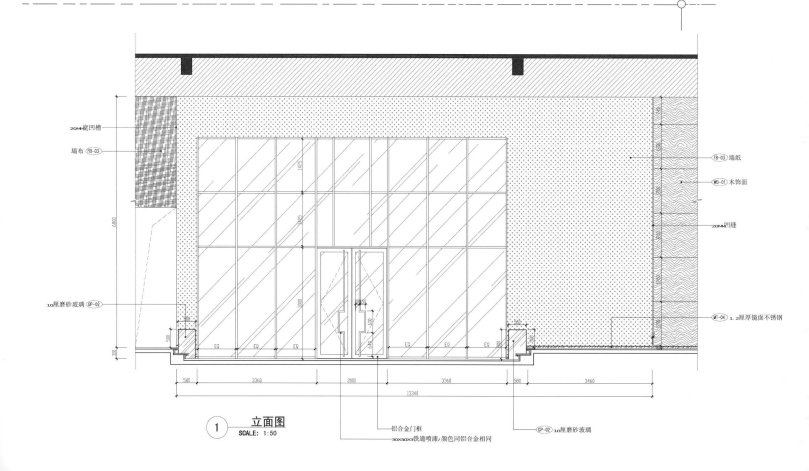

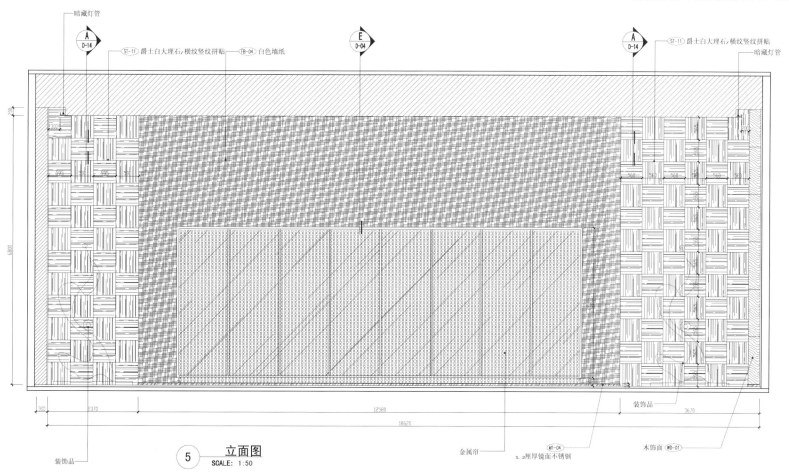

⑤ 立面图
SCALE: 1:50

Cloud View
中航城云景台会所

Design agency: Simon Chong Design Consultants Ltd.
Location: Guiyang, Guizhou
Area: 6000m²

设计单位：香港郑树芬设计事务所
项目地点：贵州贵阳
项目面积：6000 平米

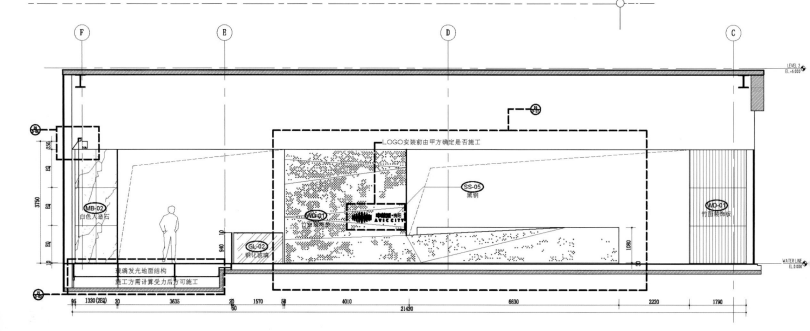

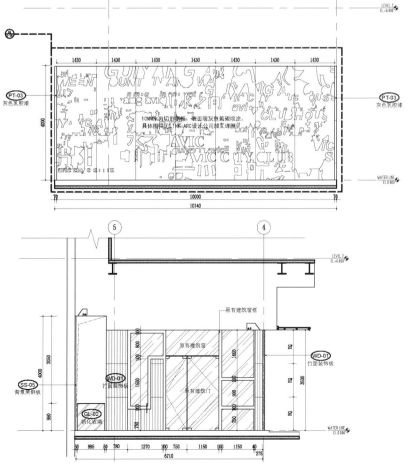

其他：书店、诊所
Other Else: Bookstore, Clinic

CITIC Books
中信书店

Design agency: SAKO Architects
Designer: Keiichiro SAKO, Hiroaki NAGATO, Jing Li
Location: Beijing
Area: 284m²
Photography: Ruijing Photo

设计单位：SAKO 建筑设计工社
设计师：迫庆一郎、长门宏明、李静
项目地点：北京
项目面积：284 平方米
摄影：锐景摄影

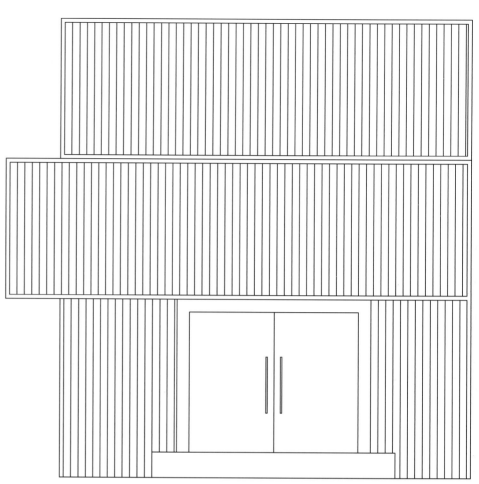

Legend Famous Car Sales Service Centre

秦皇岛市良记名车行

Design agency: Yiguan Architectural Design Studio
Designer: Chao Song
Location: Qinhuangdao, Hebei
Area: 900m²

设计单位：（香港）易观建筑设计事务所
设计师：宋超
项目地点：河北秦皇岛
项目面积：900平方米

Disney Exhibition Hall
迪士尼展厅

Design agency: HK ShengJiaYou Interior Design Co., Ltd.
Designer: Dolphin Sheng
Location: Cixi, Ningbo, Zhejiang
Area: 1200m²
Photography: Lina

设计单位：香港绳家友国际设计机构
设计师：绳家友
项目地点：浙江宁波慈溪
项目面积：1200 平方米
摄影：李影

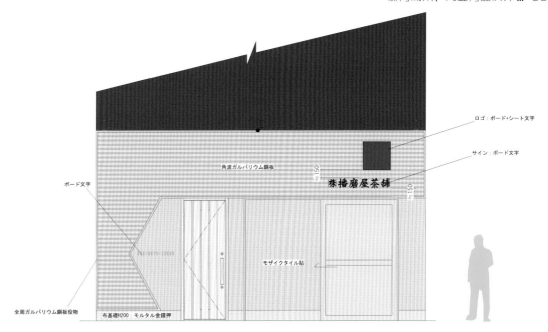

Harimaya Chaho

播磨屋茶铺

Design agency: Matsuya Art Works. Co., Ltd.
Designer: Tetsuya Matsumoto
Location: Himeji, Japan
Photography: Toshiyuki Nishimatsu

设计单位：松屋艺术设计有限公司
设计师：Tetsuya Matsumoto
项目地点：日本姬路市
摄影：Toshiyuki Nishimatsu

Careland Pharmacy

Careland 药店

Design agency: Sergio Mannino Studio, Designwajskol
Location: New York, USA

设计单位：Sergio Mannino 设计事务所、Designwajskol 设计事务所
项目地点：美国纽约

Farmacia de los Austrias
阿斯图利亚大药房

Design agency: Stone Designs
Location: Madrid

设计单位：Stone Designs 设计事务所
项目地点：马德里

Minxing Dental Clinic

民兴牙医诊所

Design agency: Urbane Design 设计单位：珥本设计
Designer: Steven Chen 设计师：陈建佑
Photography: Kevin Wu 摄影：吴启民

S.T. ENT Clinic
世典耳鼻喉科诊所

Design agency: Urbane Design
Designer: Steven Chen
Location: Taichung, Taiwan
Area: 93m²
Photography: Kevin Wu
Main materials: oak veneer, forsted glass, bianco carrara marble, graining HPL, paint glass

设计单位：珥本设计
设计师：陈建佑
项目地点：台湾台中
项目面积：93平方米
摄影：吴启民
主要材料：橡木节眼木皮、喷砂玻璃、卡拉拉白大理石、木纹美耐板、烤漆玻璃

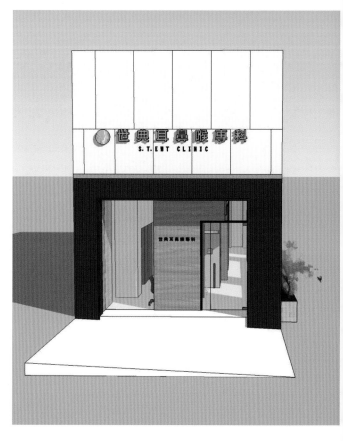

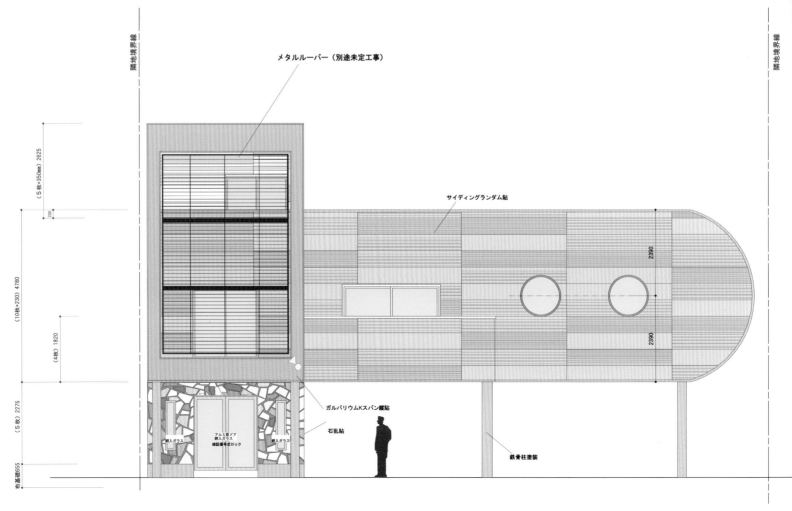

Cho clinic

长整形外科医院

Design agency: Matsuya Art Works. Co., Ltd.
Designer: Tetsuya Matsumoto
Location: Himeji, Japan
Photography: Toshiyuki Nishimatsu

设计单位：松屋艺术设计有限公司
设计师：Tetsuya Matsumoto
项目地点：日本姬路市
摄影：Toshiyuki Nishimatsu

FACE OF SHOP

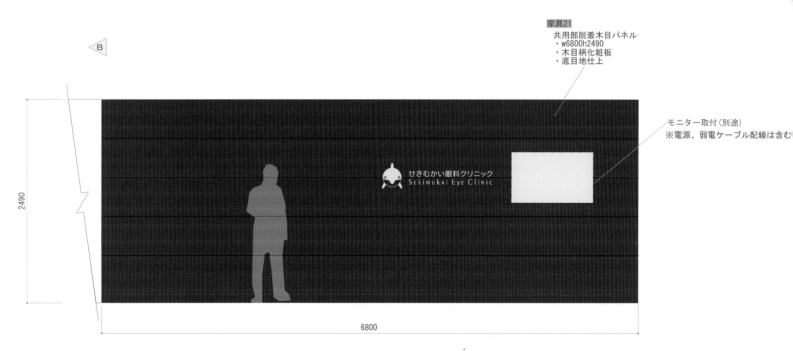

家具21
共用部脱着木目パネル
・w6800h2490
・木目柄化粧板
・底目地仕上

モニター取付（別途）
※電源、弱電ケーブル配線は含む

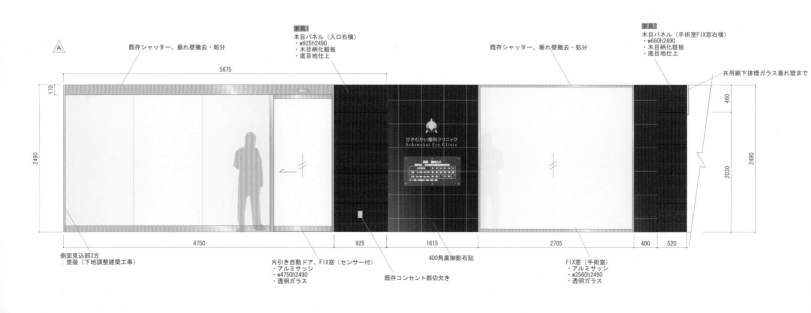

家具1
木目パネル（入口右横）
・w925h2490
・木目柄化粧板
・底目地仕上

家具2
木目パネル（手術室FIX窓右横）
・w660h2490
・木目柄化粧板
・底目地仕上

既存シャッター、垂れ壁撤去・処分

共用廊下排煙ガラス垂れ壁まで

側面見込部3方
：塗装（下地調整建築工事）

片引き自動ドア、FIX窓（センサー付）
・アルミサッシ
・w4750h2490
・透明ガラス

既存コンセント部切欠き

400角黒御影石貼

FIX窓（手術室）
・アルミサッシ
・w2560h2490
・透明ガラス

Sekimukai Eye Clinic

Sekimukai 眼科医院

Design agency: Matsuya Art Works. Co., Ltd.	设计单位：松屋艺术设计有限公司
Designer: Tetsuya Matsumoto	设计师：Tetsuya Matsumoto
Location: Nishinomiya, Japan	项目地点：日本西宫
Photography: Toshiyuki Nishimatsu	摄影：Toshiyuki Nishimatsu

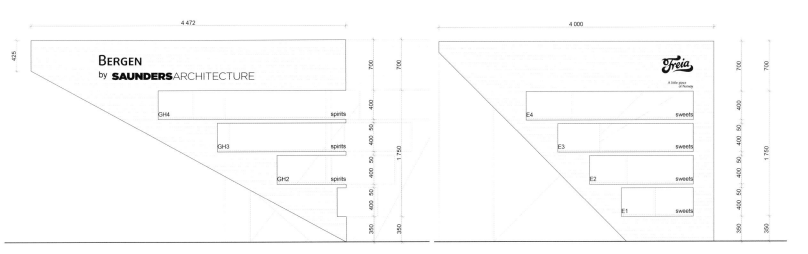

Heinemann Duty Free BGO Regionals

海恩曼免税店 BGO 特产专柜

Design agency: Saunders Architecture
Location: Bergen, Norway
Area: 50m²
Photography: Bent René Synnevag

设计单位：桑德斯建筑设计事务所
项目地点：挪威卑尔根
项目面积：50 平方米
摄影：Bent René Synnevag

Shikata Beef
志方肉工房

Design agency: Matsuya Art Works. Co., Ltd.
Designer: Tetsuya Matsumoto
Location: Kakogawa, Japan
Photography: Toshiyuki Nishimatsu

设计单位：松屋艺术设计有限公司
设计师：Tetsuya Matsumoto
项目地点：日本加古川市
摄影：Toshiyuki Nishimatsu

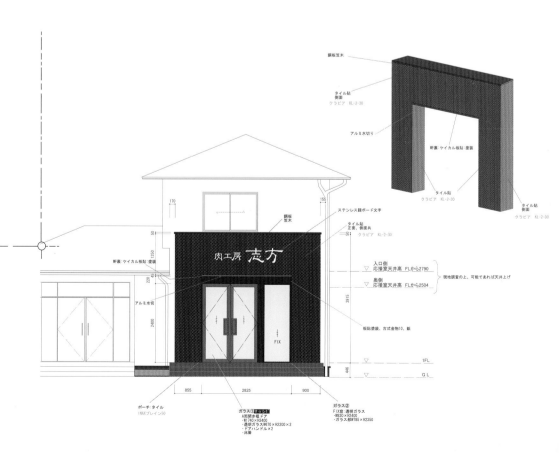

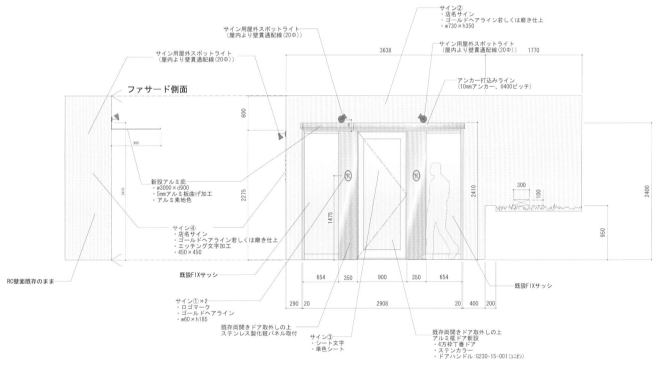

Yuka

悠卡

Design agency: Matsuya Art Works. Co., Ltd.	设计单位：松屋艺术设计有限公司
Designer: Tetsuya Matsumoto	设计师：Tetsuya Matsumoto
Location: Kobe Japan	项目地点：日本神户市
Photography: Toshiyuki Nishimatsu	摄影：Toshiyuki Nishimatsu

La Viena

维也纳

Design agency: Matsuya Art Works. Co., Ltd.
Designer: Tetsuya Matsumoto
Location: Himeji, Japan
Photography: Toshiyuki Nishimatsu

设计单位：松屋艺术设计有限公司
设计师：Tetsuya Matsumoto
项目地点：日本姬路市
摄影：Toshiyuki Nishimatsu

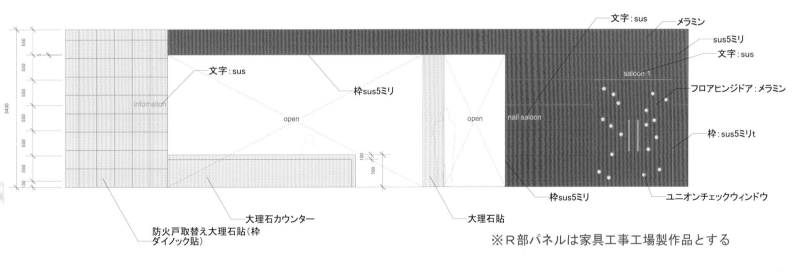

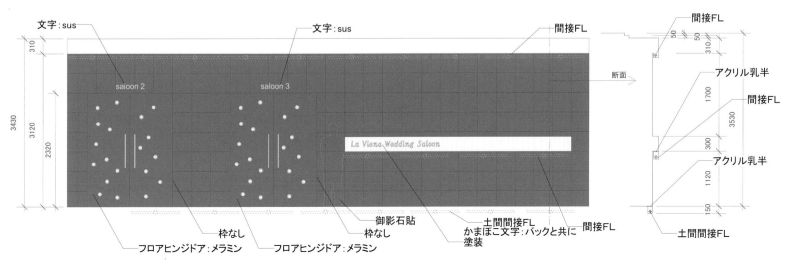

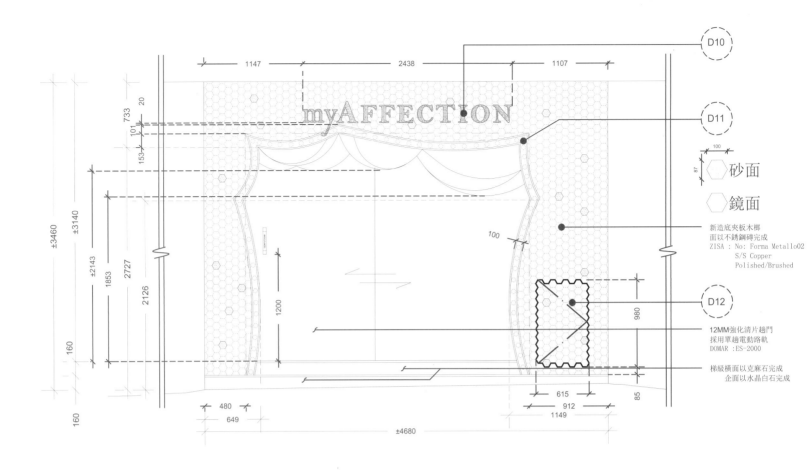

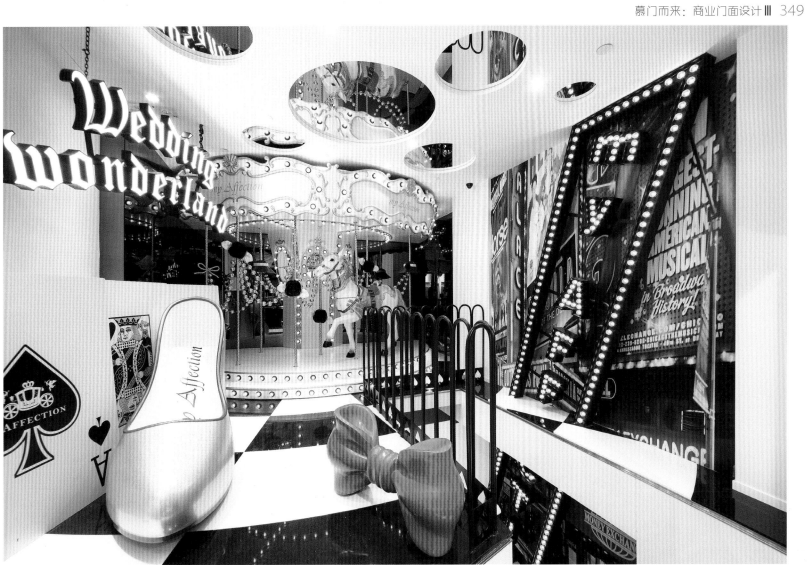

My Affection Wedding Store

囍悦一站式婚庆乐园

Design agency: Tiko Interiors Ltd.
Designer: Matthew Tong
Location: Hong Kong
Area: 450m²

设计单位：天皓设计工程有限公司
设计师：唐万强
项目地点：香港
项目面积：450 平方米

Jin Se

锦 瑟

Design agency: Fish Bone Design
Designer: Jie Na
Location: Kunming, Yunnan
Area: 2000m²
Photography: Jie Na, Jun-min Xu
Main materials: concrete, mirror stainless steel, silver travertine, aluminium plate, steel plate, preservative-treated timber

设计单位：昆明鱼骨设计事务所
设计师：纳杰
项目地点：云南昆明
项目面积：2000 平方米
摄影：纳杰、徐俊敏
主要材料：清水混凝土，镜面不锈钢，银灰洞石，铝板，钢板，防腐木

Junle Internet Café
均乐网咖

Designer: Jia-wei Shen
Location: Chengdu, Sichuan
Area: 800m²
Main materials: tile, wooden flooring, latex paint, veneer, reinforced glass, painted glass

设计师：沈嘉伟
项目地点：四川成都
项目面积：800 平方米
主要材料：地砖、木地板、乳胶漆、饰面板、钢化玻璃、喷绘玻璃

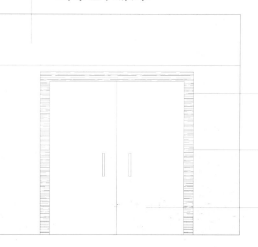

木工板基层
浅绿色乳胶漆
浅绿色乳胶漆
实木面板
玻璃

图书在版编目(CIP)数据

慕门而来：商业门面设计. 3 / 深圳市海阅通文化传播有限公司编著. -- 武汉：华中科技大学出版社, 2015.1
ISBN 978-7-5680-0615-6

I. ①慕… II.①深… III. ①商店－室内装饰设计 IV. ①TU247.2

中国版本图书馆CIP数据核字(2015)第022886号

慕门而来：商业门面设计III

深圳市海阅通文化传播有限公司 编著

出版发行：华中科技大学出版社（中国·武汉）
地　　址：武汉市武昌珞喻路1037号（邮编：430074）
出 版 人：阮海洪

责任编辑：易彩萍	责任监印：张贵君
责任校对：刘　婷	装帧设计：陈秋娣
采　　编：刘太春	

印　　刷：利丰雅高印刷（深圳）有限公司
开　　本：965 mm×1270 mm　1/16
印　　张：22
字　　数：316千字
版　　次：2015年5月第1版 第1次印刷
定　　价：368.00元 (USD 73.99)

投稿热线：(027)87545012　6365888@qq.com
本书若有印装质量问题，请向出版社营销中心调换
全国免费服务热线：400-6679-118 竭诚为您服务
版权所有　侵权必究